D1611335

❯ SPSS 14.0 Advanced Statistical Procedures Companion

Marija J. Norušis

Prentice Hall
A division of Pearson Education
1 Lake Street
Upper Saddle River, NJ 07458

For more information about SPSS® software products, please visit our Web site at *http://www.spss.com* or contact

SPSS Inc.
233 S. Wacker Dr., 11th Floor
Chicago, IL 60606-6412
Tel: (312) 651-3000
Fax: (312) 651-3668

SPSS® 14.0 Advanced Statistical Procedures Companion
Published by Prentice Hall Inc., Copyright © 2005
Upper Saddle River, NJ 07458

1 2 3 4 5 6 7 8 9 0   08 07 06 05

ISBN 0-13-174700-2

Preface

With judicious doses of common sense, well-constructed graphs, and simple statistical analyses, you can answer many interesting and important questions. That should always be your goal. However, some questions require more sophisticated statistical analyses. A statistical procedure is not like a sausage; you want to know its contents. You want to know the types of questions and data appropriate to a particular procedure. You also want to know how to interpret the output that it produces. The goal of the *SPSS 14.0 Advanced Statistical Procedures Companion* is to provide you with background information and examples for statistical procedures contained in the SPSS Advanced and Regression modules. It aims to make it less likely that you will succumb to the siren song of melodic statistical procedure names and unleash a disastrous assault on your mute data file. If you are reading this preface online go to *www.norusis.com* to view the table of contents and sample chapter for the book.

A closely related book, the *SPSS 14.0 Statistical Procedures Companion* (Norušis, 2005), focuses on the most frequently used procedures in the SPSS Base system, as well as the widely used logistic regression procedure from the Regression Models option and log linear models from the Advanced Statistics option. It also contains introductory chapters on using the software, creating and cleaning data files, testing hypotheses, and describing data. Numerous tips and warnings for analyzing data and using SPSS are incorporated. For an introduction to statistical concepts and the fundamentals of data analysis, see the *SPSS 14.0 Guide to Data Analysis* (Norušis, 2005). Sample chapters and detailed tables of contents for all three books can be found at *www.norusis.com*.

About This Book

Each of the chapters uses one or more data files, most of which can be found on the accompanying CD. Step-by-step instructions are included for obtaining the analyses described in the book. All of the statistical algorithms for the procedures can be accessed from the Help menu in the software. Choose Help > Statistical Algorithms.

Acknowledgments

I thank SPSS users who have read my books for many years. Their suggestions and encouragement are much appreciated. Please continue to send comments and suggestions about my books to *marija@norusis.com*. Please send comments or suggestions about the software to *suggest@spss.com*.

I also wish to thank the SPSS staff who has assisted in various ways in the preparation of this book. In particular, I am indebted to David Nichols for his many substantive suggestions and corrections. I am also grateful to Yvonne Smith, JoAnn Ziebarth, Kris Horgen, and David O'Neil for their superb editorial and production support. Finally, I wish to thank my barista, Bruce Stephenson.

Marija Norušis

Contents

3   *Multinomial Logistic Regression*   *43*

4   *Ordinal Regression*   69

8   *Cox Regression*   *135*

9   *Variance Components*   *173*

10   *Linear Mixed Models*   *197*

11 *Nonlinear Regression* 247

Model Selection Loglinear Analysis

You are often interested in examining relationships between categorical variables. Are happiness, good health, a substantial bank account, and the right partner (or perhaps no partner at all) related? What about infant survival, cigarette smoking, maternal age, and prenatal care? You can create crosstabulations of pairs of variables, controlling for the rest, and attempt to examine the relationships, but before long you'll have too many tables with too few cases in each to draw any reliable conclusions.

Loglinear modeling is a statistical technique that is often used for analyzing multidimensional crosstabulations. Instead of examining numerous individual crosstabulations, you systematically examine all orders of interactions among variables. You identify interactions that are significantly different from 0 and build a model that includes them. Loglinear models are similar to multiple regression models. In loglinear models, all variables that are used for classification are independent, or predictor, variables, and the dependent variable is the log of the number of cases in a cell of the multiway crosstabulation.

SPSS has three procedures for modeling multiway crosstabulation data. The procedures differ in important ways.

- Use the Model Selection Loglinear Analysis procedure if you have few ideas in advance about the model that you want to fit. The Model Selection procedure automatically screens a large number of models for you. This capability is important because the number of possible models is large, even for a small number of variables. The Model Selection procedure is restricted to hierarchical loglinear models; parameter estimates are calculated only for saturated models.

- Use the General Loglinear Analysis procedure to estimate parameters of any loglinear model, including those with observed and/or predicted cell counts of 0.

- Use the Logit Loglinear Analysis procedure to estimate the parameters of a model when one of the variables is considered to be the dependent variable. (For more information, see Chapter 2.)

Loglinear Modeling Basics

Loglinear modeling is based on the idea of summarizing a crosstabulation in terms of main effects and interaction parameters, much like the general linear model for analysis of variance. You don't single out one of the variables as *dependent*; rather, the dependent variable is the log of the number of cases in each cell of the crosstabulation. The strategy is to determine how many of these interactions are needed to represent the data adequately and then to interpret the interactions so that you can understand the underlying relationships among the variables.

Before you start evaluating models, you must select the categorical variables that you want to study. Although you may be tempted to include as many variables as you can collect, this is often a poor strategy because the number of cells in a crosstabulation rapidly increases as you add variables. If you have a large number of cells with few observations, you can neither estimate the parameters well nor assess the goodness of fit of the model well.

To use loglinear models, all observations in a crosstabulation must be independent. That is, each case can appear only once in the table. For more information about the sampling models for categorical data, see the "General Loglinear Analysis" chapter in the *SPSS 14.0 Statistical Procedures Companion* (Norušis, 2005).

A Two-Way Table

As an introduction to the basics of fitting loglinear models, consider the two-way table of marital status and happiness shown in Figure 1-1. The data are from the General Social Survey (Davis and Smith, 1972–2002). With two variables, the loglinear equations are uncluttered and easy to understand.

Figure 1-1
Crosstabulation of marital status and happiness

| | | | General happiness | | Total |
			Happy	Not happy	
Marital status	Married	Count	566	38	604
		% within Marital status	93.7%	6.3%	100%
	Split	Count	320	72	392
		% within Marital status	81.6%	18.4%	100%
	Never married	Count	313	60	373
		% within Marital status	83.9%	16.1%	100%
Total		Count	1199	170	1369
		% within Marital status	87.6%	12.4%	100%

The crosstabulation shows that there appears to be a relationship between happiness and marital status. Almost 94% of married people claim that they are happy compared to only 82% of people who are divorced, widowed, or separated. A chi-square test of independence leads you to reject the null hypothesis that the two variables are independent. When you test the null hypothesis that the two variables are independent, using the familiar chi-square test of independence, you are fitting a simple loglinear model.

The Saturated Model

The components that you can include in a loglinear model are the same as those in a factorial analysis of variance model—an intercept, main effects, and interactions of all orders. In this example, you have two main effects—marital status and happiness—and one two-way interaction—the interaction of marital status and happiness. A model that includes all possible main effects and interactions is called a **saturated model**. By definition, a saturated model exactly reproduces the observed counts in each cell. The dependent variable is the natural log of the number of cases in each cell of the table. (The natural log of a number is the power to which the number e is raised to give that number. For example, the natural log of the first cell entry is 6.339, since $e^{6.339} = 566$.)

You can express the log of the observed cell counts in each cell in Figure 1-1 as a linear combination of parameters:

$$\log(566) = \mu + \hat{\lambda}^{happy} + \hat{\lambda}^{married} + \hat{\lambda}^{happy\,married}$$

$$\log(38) = \mu + \hat{\lambda}^{unhappy} + \hat{\lambda}^{married} + \hat{\lambda}^{unhappy\,married}$$

$$\log(320) = \mu + \hat{\lambda}^{happy} + \hat{\lambda}^{split} + \hat{\lambda}^{happy\,split}$$

$$\log(72) = \mu + \hat{\lambda}^{unhappy} + \hat{\lambda}^{split} + \hat{\lambda}^{unhappy\,split}$$

$$\log(313) = \mu + \hat{\lambda}^{happy} + \hat{\lambda}^{single} + \hat{\lambda}^{happy\,single}$$

$$\log(60) = \mu + \hat{\lambda}^{unhappy} + \hat{\lambda}^{single} + \hat{\lambda}^{unhappy\,single}$$

For a saturated model, the estimate for μ is simply the average of the logs of the frequencies in all table cells. Estimates for the lambda parameters are obtained in a manner similar to that for analysis of variance.

Main Effects

Each individual category of the row and column variables has an associated lambda parameter (λ). In this example, there is a lambda for being married, a lambda for being split up, and a lambda for never being married. The term $\lambda^{married}$ is the effect of being in the *married* category of the marital status variable. The main effect *happy* has two lambda parameters, since it has two categories—*happy* and *unhappy*.

For a saturated model, the values of the main-effect parameters depend only on the number of cases in each of the categories. If most people in the study are happy, the lambda parameter for *happy* will be large and positive. If few people are single, compared to married and split, the lambda parameter for *single* will be negative. For a saturated model, the main-effect parameters don't tell you anything about the relationships between variables; they tell you only about the distribution of cases within the categories of a particular variable.

The main-effect parameters for a saturated model are calculated by finding the difference between the grand mean of the log cell counts in the table (the intercept) and the mean of the log cell counts for a particular row or column. For example, the parameter estimate for *happy* is

$$\hat{\lambda}^{happy} = 5.95 - 4.98 = 0.97$$

where 4.98 is the average of the logs of the cell counts in the entire table (the intercept) and 5.95 is the average of the logs of the cell counts in the *Happy* column. Similarly,

$$\hat{\lambda}^{\text{married}} = 4.99 - 4.98 = 0.01$$

Interactions

In most analyses, you are interested in the interactions between variables. You want to know whether certain combinations of values of variables are more or less likely than you would expect, based on the values of the individual variables. For example, the term $\lambda^{\text{happymarried}}$ represents the interaction of being happy and being married. The interaction parameters indicate how much difference there is between the sums of the effects of the variables taken individually and collectively. They represent the "boost" or "interference" associated with particular combinations of the values. For example, if marriage increases happiness, the number of cases in the *happy* and *married* cell would be larger than the number you would expect based on only the frequency of married people (λ^{married}) and the frequency of happy people (λ^{happy}). This excess would be represented by a positive value for $\lambda^{\text{happymarried}}$. If marriage decreases happiness, the value for the interaction parameter would be negative. If marriage neither increases nor decreases happiness, the interaction parameter would be close to 0.

You can have any number of variables in an interaction. The number of lambda parameters that represent the interaction is the product of the number of categories of the variables involved in the interaction. If you have three categories of marital status and two categories of happiness, there are six lambda parameters for the interaction of marital status and happiness.

To calculate interaction parameters for a table with two categorical variables, you find the difference between the log of the observed cell count in a cell and the log of the predicted cell count, based on only the intercept and the main effects. For example, the estimated interaction parameter for the *happy* and *married* cell is

$$\hat{\lambda}^{\text{happymarried}} = \log(566) - (\mu + \hat{\lambda}^{\text{happy}} + \hat{\lambda}^{\text{married}})$$
$$\hat{\lambda}^{\text{happymarried}} = 6.34 - (4.98 + 0.97 + 0.01) = 0.38$$

Constraints on the Parameters

In the happiness and marital status example, there are six cells in the crosstabulation. There are 12 parameters in the loglinear model (the constant, three parameters for marital status, two for happiness, and six for the interaction of marital status and happiness.) To uniquely estimate the lambda parameters, you need to impose certain constraints on them. In the SPSS Model Selection Loglinear Analysis procedure, the constraint is that the lambdas must sum to 0 across categories of a variable.

Examining Parameters in a Saturated Model

To obtain parameter estimates for the table shown in Figure 1-1, open the data file *gsslogmodel.sav*, and from the menus choose:

Analyze
  Loglinear
    Model Selection...

▶ Factor(s): happy2(1,2), marital3(1,3)

Define Range for each variable (select variable and click Define Range)

Model Building
  ⊙ Enter in single step

Model...
    Specify Model
    ⊙ Saturated

Options...
    Display for Saturated Model
    ☑ Parameter estimates
    Model Criteria
    Delta: 0

Note that in the Options dialog box, you set the delta criterion to 0. This prevents SPSS from adding the default 0.5 to all observed cell counts and maintains the original observed counts for a saturated model.

The parameter estimates are shown in Figure 1-2. There are estimates for the two-way interaction *happy2*marital3* and for the main effects (*happy2* and *marital3*). The values for *happy2* and the first interaction parameter should be familiar, since you calculated them previously. The number of estimates displayed for each effect depends on the number of categories for the variables that compose it. For main effects, the number of parameter estimates is one less than the number of categories of the variable, which is why one parameter estimate is displayed for *happy2* and two are displayed for *marital3*.

Figure 1-2

Parameter estimates

				Estimate	Std. Error	Z	Sig.	95% Confidence Interval	
								Lower Bound	Upper Bound
Effect	happy2*marital3	Parameter	1	.376	.064	5.847	.000	.250	.503
			2	-.228	.057	-4.021	.000	-.340	-.117
	happy2	Parameter	1	.974	.042	22.932	.000	.891	1.057
	marital3	Parameter	1	.011	.064	.173	.863	-.115	.137
			2	.046	.057	.802	.422	-.066	.157

For interactions, the number of parameter estimates displayed is the product of one less than the number of categories for each of the individual variables involved in the interaction. The *happy2*marital3* interaction displays two parameter estimates: the product of 2 (one less than the number of marital categories) and 1 (one less than the number of happiness categories). For each interaction in the model, the interaction parameters are numbered sequentially, just like the main effects. To determine the correspondence between the variable categories involved in an interaction and the parameter numbers, use the following rules:

- The order of the variable names listed for the interaction determines the sequence in which parameter estimates are displayed. The levels of the first variable increment the most slowly, and the levels of the last variable increment the most quickly.

- Parameter estimates that involve the last category of any variable in the interaction are not displayed.

In Figure 1-2, the two-way interaction *happy2*marital3* shows two parameter estimates. The first one is for the interactions of the first level of *happy2* (happy) by the first level of *marital3* (married); the second is for the first level of *happy2* (happy) by the second level of *marital3* (split). Parameter estimates that involve the last (in this case, second) level of *happy2* and the last (in this case, third) level of *marital3* are not displayed.

Consider how the interaction parameters are displayed if the three-way interaction *happy2*marital3*income3* is in the model. (The last character in the variable name tells you how many levels each variable has.) The sequence in which the parameter estimates are displayed is shown in Table 1-1.

Table 1-1
Sequence of interaction parameters

Parameter	Happy2	Marital3	Income3
1	1 (happy)	1 (married)	1 (bottom third)
2	1 (happy)	1 (married)	2 (middle third)
3	1 (happy)	2 (split)	1 (bottom third)
4	1 (happy)	2 (split)	2 (middle third)

Calculating the Missing Parameter Estimates

If you want to examine all parameter estimates and not only those displayed, you have two choices. Since the parameter estimates must sum to 0 over all categories of a variable, you can calculate the missing parameter estimates by adding up all of the estimates for a category of a variable and then setting the missing one to the negative of that value. Or, you can recode the variables so that another category is the last one and obtain the missing value from the new analysis.

Main-Effects Parameters

The parameter estimates for each main effect are listed sequentially in the output, so you can calculate the missing estimates by summing the values of all of the displayed parameters. For example, the coefficient shown for *happy2* is 0.97, which is the coefficient corresponding to the response *happy*. The coefficient corresponding to the response *not happy* must be –0.97.

It's a little more work when you have more than one parameter estimate for a main effect, since you must add up all of the estimates that are displayed. For example, to get the parameter estimate for *never married* (the third category of the variable *marital3*), add up all of the displayed estimates, which equals 0.057. To get a sum of 0 over all categories of the variable, the estimate for *never married* is –0.057. (This is the number displayed for *never married* if you use the Recode facility to make another category the last [omitted] category.)

Interaction Parameters

It's easy to get the values for the missing main-effect parameter, since there is only one missing for each main effect. For the interaction parameters, it is more complicated because the number you have to fill in depends on the product of the number of categories of each variable involved in the interaction.

For the two-way interaction *happy2*marital3*, there are six parameter estimates (2×3). However, in Figure 1-2, you see only two—those for *happy2* $= 1$ * *marital3* $= 1$ and *happy2* $= 1$ * *marital3* $= 2$. The missing ones involve the last categories of the two variables. You can create a table such as that shown in Table 1-2 and use it to calculate the missing values.

In Table 1-2, the reported interactions and the marginal sums of 0 are shown in bold. You have to calculate the missing elements that make the row and column sums equal to 0.

Table 1-2
Missing interactions

happy2	*marital3*			Sum
	1 (married)	2 (split)	3 (never married)	**0**
1 (happy)	**0.376**	**−0.228**	−0.148	**0**
2 (not happy)	−0.376	0.228	0.148	**0**
Sum	**0**	**0**	**0**	**0**

Testing Hypotheses about Parameters

You can test whether each of the parameters in the loglinear model is significantly different from 0, based on the statistics shown in Figure 1-2. For sufficiently large sample sizes, the test of the null hypothesis that lambda is 0 can be based on this Z value, since the standardized lambda is approximately normally distributed with a mean of 0 and a standard deviation of 1 if the model fits the data. Lambdas with Z values greater than 1.96 in absolute value can be considered significant at the 0.05 level. When tests for many lambdas are calculated, however, the usual problem of multiple comparisons arises. That is, when many comparisons are made, the probability that some are found to be significant when there is no effect increases rapidly.

Confidence intervals for each parameter are constructed in the usual manner. To obtain the 95% confidence interval, you add and subtract 1.96 times the standard error from the parameter estimate. If the 95% confidence interval does not contain 0, you can reject the null hypothesis that the parameter value is 0 in the population.

Based on the statistics shown in Figure 1-2, you can reject the null hypothesis that the individual interaction parameters are 0. This tells you that there appears to be a relationship between marital status and reported happiness. The parameters for the main effect *happy2* are significantly different from 0 because there are many more people who are happy than unhappy. The main-effect parameters for marital status are not significantly different from 0, since the number of people in each of the three categories is more similar.

Fitting an Independence Model

Representing an observed-frequency table with a loglinear model that contains as many parameters as there are cells (a saturated model) is a good starting point for exploring other models that could be used to represent the data. Parameters that have small values might be excluded from subsequent models.

To illustrate the general procedure for fitting a model that does not contain all possible parameters (an unsaturated or custom model), consider the familiar independence hypothesis for a two-way table. If variables are independent, they can be represented by a loglinear model that does not have any interaction terms. For example, if happiness and marital status are independent,

$$\log(m_{ij}) = \mu + \lambda_i^{happy} + \lambda_j^{marital}$$

Note that m_{ij} is no longer the observed frequency in the (i,j)th cell, but it is now the expected frequency based on the model. Only saturated models necessarily reproduce the observed data.

Specifying the Model

To fit an independence model, from the menus choose:

Analyze
  Loglinear
    Model Selection...

▶ Factor(s): happy2(1,2), marital3(1,3)

Define Range for each variable (select variable and click Define Range)

Model Building
⊙ Enter in single step

Model...
  Specify Model
  ⊙ Custom
      ▶ happy2, marital3

Options...
  Display
  ☑ Frequencies
  ☑ Residuals

Checking Convergence

When you fit an unsaturated model, the solution is obtained using an iterative algorithm. Each time an estimate is obtained, it is called an **iteration**. You can set the maximum number of iterations that the algorithm performs. You can also control the largest amount by which successive estimates of expected cell counts may differ. This is called the **convergence criterion**. (The default convergence criterion is the larger of 0.25 and 0.001 times the largest observed cell count.) If the convergence criterion is not reached before the maximum number of iterations, the algorithm stops without converging.

The table in Figure 1-3 shows the number of iterations required for convergence and the criterion used for convergence. For this example, two iterations were required for convergence. You should always check to make sure that the solution converged. If not, increase the number of iterations.

Figure 1-3
Convergence information

Generating Class	happy2, marital3
Number of Iterations	2
Max. Difference between Observed and Fitted Marginals	.000
Convergence Criterion	.566

Observed and Expected Counts

When you fit a saturated model, all observed and expected cell counts are equal because you have as many parameters as you do cells in the table. When you fit an unsaturated model, you don't expect to see a perfect fit. Figure 1-4 contains the cell counts and residuals from the independence model. (The expected values are identical to those obtained from the usual formulas for expected values in a two-way crosstabulation.)

The columns labeled *Observed* and *Expected* contain the observed cell counts and the estimated cell counts based on the independence model. Each count is also given as a percentage of the total count. The difference between the observed and expected counts is shown in the *Residuals* column. For example, the observed number of happy married people is 566, while the number predicted by the independence model is 529. The residual is 37 ($566 - 529$).

As in regression analysis, it's useful to standardize the residuals so that unusually large values are easy to identify. In this case, the square root of the expected cell count is used to standardize the residuals. For example, the standardized residual for the first cell in Figure 1-4 is

$$\frac{37}{\sqrt{529}} = 1.6$$

This value is displayed in the column labeled *Std Residuals*. If the model is adequate, the standardized residuals are approximately normally distributed with a mean of 0 and a standard deviation of less than 1. Standardized residuals greater than 1.96 or less than -1.96 suggest discrepancies that should be examined, since they are unlikely to occur if the model is adequate. Particular combinations of cells with large standardized residuals may suggest which other models might be more appropriate. The same types of diagnostics for residuals used in regression analysis can be used in loglinear models.

Figure 1-4

Cell counts and residuals

				Observed		Expected			
				Count	%	Count	%	Residuals	Std. Residuals
happy2	Happy	marital3	Married	566.000	41.3%	528.996	38.6%	37.004	1.609
			Split	320.000	23.4%	343.322	25.1%	-23.322	-1.259
			Never married	313.000	22.9%	326.682	23.9%	-13.682	-.757
	Not happy	marital3	Married	38.000	2.8%	75.004	5.5%	-37.004	-4.273
			Split	72.000	5.3%	48.678	3.6%	23.322	3.343
			Never married	60.000	4.4%	46.318	3.4%	13.682	2.010

Chi-Square Goodness-of-Fit Tests

The test of the hypothesis that a particular model fits the observed data can be based on the familiar **Pearson chi-square statistic**, which is calculated as

$$\chi^2 = \sum_i \frac{(O_i - E_i)^2}{E_i}$$

where the subscript *i* includes all cells in the table. An alternative statistic is the **likelihood-ratio chi-square**, which is calculated as

$$G^2 = 2\sum_i O_i \ln\left(\frac{O_i}{E_i}\right)$$

For large sample sizes, these statistics are equivalent. The advantage of the likelihood-ratio chi-square statistic is that it, like the total sums of squares in analysis of variance, can be subdivided into interpretable parts that add up to the total.

Figure 1-5 shows that, for the independence model, the value of the Pearson chi-square statistic is 40.48 and the likelihood-ratio chi-square is 38.22. The degrees of freedom for a model are the number of cells in the table minus the number of independent parameters in the model. In this example, there are six cells and four independent parameters to be estimated (the grand mean, one parameter for happiness, and two parameters for marital status), so there are two degrees of freedom. (There are only four independent parameters because of the constraint that parameter estimates must sum to 0 over the categories of a variable. Therefore, the value for one of the categories is determined by the values of the others and is not, in a statistical sense, independent.)

Because the observed significance level associated with both chi-square statistics is very small (less than 0.0005), the independence model is rejected. Note that both the Pearson and likelihood-ratio chi-square statistics displayed for the independence model are the same as the chi-square value displayed by the SPSS Crosstabs procedure.

Figure 1-5
Chi-square goodness-of-fit test

	Chi-Square	df	Sig.
Likelihood Ratio	40.480	2	.000
Pearson	38.217	2	.000

Hierarchical Models

A saturated loglinear model contains all possible effects. For example, a saturated model for a two-way table contains terms for the row main effects, the column main effects, and their interaction. Different models can be obtained by deleting terms from a saturated model. The independence model is derived by deleting the interaction effect. Attention is often focused on a special class of models called **hierarchical models**.

In a hierarchical model, if a term exists for the interaction of a set of variables, there must be lower-order terms for all possible combinations of these variables. For a two-variable model, this means that the interaction term can be included only if both main effects are present. For a three-variable model, if the term λ^{ABC} is included in a model, the terms λ^A, λ^B, λ^C, λ^{AB}, λ^{BC}, and λ^{AC} must also be included.

Generating Classes

To describe a hierarchical model, it is sufficient to list the highest-order terms in which variables appear. This is called the **generating class** of a model. For example, the specification A*B*C indicates that a model contains the term λ^{ABC} and all of its lower-order relations. (Terms are "relatives" if all variables that are included in one term are also included in the other. For example, the term λ^{ABCD} is a higher-order relative of the terms λ^{ABC}, λ^{BCD}, λ^{ACD}, and λ^{ABD}, as well as all other lower-order terms involving variables A, B, C, or D. Similarly, λ^{AB} is a lower-order relative of both λ^{ABC} and λ^{ABD}.) The model

$$\log(m_{ijk}) = \mu + \lambda_i^A + \lambda_j^B + \lambda_k^C + \lambda_{ij}^{AB}$$

can be represented by the generating class (A*B)(C), since AB is the highest-order term in which A and B occur, and C is included in the model only as a main effect.

Selecting a Model

Even if you restrict your attention to hierarchical models, many different models are possible for a set of variables. How do you choose among them? The same guidelines discussed for model selection in regression analysis apply to loglinear models. (For more information about regression, see the *SPSS 14.0 Statistical Procedures Companion* [Norušis, 2005]). A model should fit the data and be substantively interpretable and as simple (parsimonious) as possible. For example, if a model without higher-order interaction terms fits the data almost as well as a model with higher-order interactions, the simpler model is usually better, since higher-order interaction terms are difficult to interpret.

Evaluating Interactions

One strategy for selecting a model is to look not at individual terms but to examine the contribution of *all* interaction terms of a particular order. For example, you can test the contribution of all three-way interactions or all three-way and higher interactions. The mechanics are similar to those in linear regression when you test the contribution of variables added (or removed) from a model.

In regression analysis, the change in multiple R^2 when a variable is added to a model indicates the additional information conveyed by the variable. In loglinear analysis, the decrease in the value of the likelihood-ratio chi-square statistic when terms are added to the model indicates their contribution to the model. (Remember that R^2 increases when additional variables are added to a model, since large values of R^2 [and small observed significance levels] are associated with good models. Chi-square values decrease when terms are added, since small values of chi-square [and large observed significance levels] are associated with good models.)

As an example of evaluating the effects of interactions, consider the happiness and marital status data when two additional variables, total income category (*income3*) and gender (*sex2*) are also included. You now have:

- Four main effects: happiness, marital status, income, and gender

- Six two-way interactions: happiness-by-marital status, happiness-by-income, happiness-by-gender, marital status-by-income, marital status-by-gender, and income-by-gender

- Four three-way interactions: happiness-by-marital status-by-income, happiness-by-marital status-by-gender, marital status-by-income-by-gender, and happiness-by-income-by-gender

- One four-way interaction: happiness-by-marital status-by-income-by-gender

To study the interaction effects, from the menus choose:

Analyze
  Loglinear
    Model Selection...

▶ Factor(s): happy2(1,2), marital3(1,3), income3(1,3), sex2(1,2)

Define Range for each variable (select variable and click Define Range)

Model Building
⊙ Enter in single step

Model...
    Specify Model
    ⊙ Saturated

Options...
    Display for Saturated Model
    ☑ Parameter estimates
    ☑ Association table
    Model Criteria
    Delta: 0

Tests for Interactions

The SPSS Model Selection Loglinear Analysis procedure calculates three related tests for interaction terms:

- Tests that all kth-order interaction effects are 0
- Tests that all kth- and higher-order interaction effects are 0
- Tests that individual interactions are 0

All of the tests convey slightly different information about the model.

Testing that Kth-Order Interactions Are Zero

To test whether interaction terms of a particular order are 0, you fit the model with and without the interactions and look at the change in the likelihood-ratio chi-square. In general, the test of the hypothesis that the kth-order terms are 0 is based on

$$\chi^2 = \chi_{k-1}^2 - \chi_k^2$$

where χ_k^2 is the value for the model that includes all interactions of order k and lower, and χ_{k-1}^2 is the chi-square value for the model without the kth-order effects. The degrees of freedom for the test is the difference in degrees of freedom between the two models.

For example, if you want to test whether all third-order interaction terms are 0, you must fit two models: the model with a constant, main effects, two-way interactions, and three-way interactions and the same model excluding the three-way interactions. Then you compute the change in the likelihood-ratio chi-square between the two models. The degrees of freedom for the change are the degrees of freedom associated with the three-way interactions. The SPSS Model Selection Loglinear Analysis procedure systematically adds interaction terms to the model and reports the changes in chi-square values.

Figure 1-6 shows the change in the likelihood-ratio chi-square as interaction terms are added to a model that contains all lower-order interactions. If you remove main effects ($k = 1$) from a model, leaving only the intercept, the change in the likelihood-ratio chi-square is 878. The small observed significance level for the change tells you to reject the null hypothesis that the main-effect terms are 0. You need the main effects in the model. Similarly, you can't reject the null hypothesis that second-order interactions are 0, since adding them to a model with main effects results in a large change in the chi-square value and a small observed significance level for the change. However, you cannot reject the null hypothesis that third-order interactions are 0, based on the large observed significance level for $k = 3$. When you add third-order interactions to a model with main effects and second-order interactions, the fit of the model does not improve much. You also cannot reject the null hypothesis that the fourth-order interaction is 0, since adding it to the model results in a nonsignificant change in the chi-square value.

Figure 1-6
Tests that k-way effects are 0

			df	Likelihood Ratio		Pearson		Number of Iterations
				Chi-Square	Sig.	Chi-Square	Sig.	
K-way Effects[2]	K	1	6	878.484	.000	952.009	.000	0
		2	13	341.687	.000	379.931	.000	0
		3	12	15.877	.197	15.485	.216	0
		4	4	3.839	.428	3.767	.438	0

1. Tests that k-way and higher order effects are zero.

2. Tests that k-way effects are zero.

Testing that Kth-Order and Higher Interactions Are Zero

In a hierarchical model, you must have all lower-order interactions that correspond to higher-order interactions. For example, if you include all three-way interactions, you must include all two-way interactions as well. This is because, by definition, a higher-order interaction implies the existence of all lower-order interactions and main effects. If you exclude third-order interactions, you have to exclude fourth-order interactions as well to maintain the hierarchical nature of the model.

Figure 1-7 contains tests of the hypothesis that kth-order and higher interactions are 0. You look at the highest-order interaction first ($k = 4$). Based on the large observed significance level, you cannot reject the null hypothesis that the fourth-order interaction is 0. Since 4 is the highest-order interaction, the tests shown in Figure 1-6 and Figure 1-7 are identical. There is no "and higher" for fourth-order interactions.

Figure 1-7
Tests that k-way and higher-order effects are 0

		df	Likelihood Ratio		Pearson		Number of Iterations
			Chi-Square	Sig.	Chi-Square	Sig.	
K-way and Higher Order Effects	K	35	1239.887	.000	1351.191	.000	0
		29	361.403	.000	399.183	.000	2
		16	19.716	.233	19.252	.256	5
		4	3.839	.428	3.767	.438	3

1. Tests that k-way and higher order effects are zero.

2. Tests that k-way effects are zero.

The third row is a test of the null hypothesis that all three-way and higher interactions are 0. The observed significance level is large ($p = 0.26$), so you can't reject the null hypothesis that three- and four-way interactions are 0. Since the observed significance level for testing that two-way and higher interactions are 0 is small, you know that you must include two-way interactions. Using the table shown in Figure 1-7, you want to find the first row that has a reasonably large observed significance level, indicating that interactions of that order and higher are not significantly different from 0. Then you can fit the model with all lower-order interactions.

Likelihood-ratio chi-square values can be added together, so you can derive Figure 1-7 from Figure 1-6. For example, the test that third- and fourth-order interactions are 0 is the sum of the changes in chi-square values when the third-order interactions are removed from the model and the change in the chi-square value when the fourth-order interaction is removed. From Figure 1-6, the chi-square change value for the third-order interactions is 15.877, and the change for the fourth-order interaction is 3.839. The sum of these two values, 19.716, is the entry (in Figure 1-7) for the test of three-way and higher interactions.

Testing Individual Terms in the Model

The two tests described in the previous section provide an indication of the collective importance of effects of various orders. They do not, however, test the individual terms. That is, although the overall hypothesis that second-order terms are 0 may be rejected, this does not mean that every second-order effect is present.

One strategy for testing individual terms is to fit two models that differ only in the presence of the effect to be tested. Such models are called **nested**. The difference between the two likelihood-ratio chi-square values, sometimes called the **partial chi-square**, also has a chi-square distribution and can be used to test the hypothesis that the effect is 0. For example, to test that the *happy2*marital3*income3* effect is 0, you fit a model with all three-way interactions. Then you remove the *happy2*marital3*income3* interaction and look at the change in the likelihood-ratio chi-square. If the observed significance level for the change is large, you can conclude that the interaction parameter is not significantly different from 0. If the change in the chi-square value is large and the observed significance level is small, you conclude that the effect has a non-zero coefficient. The tests of partial association for two-way interactions are computed by comparing a model with all two-way interactions (and no higher-order interactions) to models in which each of the two-way interactions is removed in turn.

Figure 1-8 contains the partial chi-square values and their observed significance levels for all effects in the *happy2*marital3*income3*sex2* table. The last column indicates the number of iterations required to achieve convergence. Note that the observed significance levels are large for all three-way effects except for *marital3*income3*sex2*.

Figure 1-8
Tests of partial association

		df	Partial Chi-Square	Sig.	Number of Iterations
Effect	happy2*marital3*income3	4	3.362	.499	5
	happy2*marital3*sex2	2	.244	.885	3
	happy2*income3*sex2	2	.697	.706	3
	marital3*income3*sex2	4	11.484	.022	4
	happy2*marital3	2	22.848	.000	5
	happy2*income3	2	12.352	.002	5
	marital3*income3	4	207.609	.000	4
	happy2*sex2	1	2.472	.116	5
	marital3*sex2	2	11.944	.003	5
	income3*sex2	2	9.776	.008	5
	happy2	1	789.052	.000	2
	marital3	2	69.410	.000	2
	income3	2	17.894	.000	2
	sex2	1	2.127	.145	2

Model Selection Using Backward Elimination

As in regression analysis, another way to try to arrive at a "best" model is by using variable-selection algorithms. Forward selection adds effects to a model, while backward elimination starts with all effects in a model and then removes those that do not satisfy the criterion for remaining in the model. Since backward elimination appears to be the better procedure for model selection in hierarchical loglinear models, it is the only method available in the SPSS Model Selection Loglinear Analysis procedure.

The initial model for backward elimination need not be saturated but can be any hierarchical model that you define in the Loglinear Analysis Model dialog box. At the first step, the effect whose removal results in the least-significant change in the likelihood-ratio chi-square is eliminated, provided that the observed significance level is larger than the criterion for remaining in the model. (As in regression, you remove variables that have large observed significance levels for the change in the statistic of interest.) To ensure a hierarchical model, only effects corresponding to the generating class are eligible to be removed at a given step. For example, if the generating class is *happy2*marital3*income3*sex2*, the first step examines only the fourth-order interaction.

To build a model using backward elimination, recall the Model Selection Loglinear Analysis dialog box and choose:

Model building
⊙ Use backward elimination

Figure 1-9 shows output for step 0. The model is identified by its generating class. The chi-square value is 0, since the initial model is saturated. At each step, all eligible effects are considered for removal from the model, while maintaining the hierarchical nature of the model. The rows labeled *Deleted Effect* at each step show the change in the chi-square value if each of the eligible effects is removed. Elimination of the fourth-order interaction results in a chi-square change of 3.84, which has an associated significance level of 0.43. Since this significance level is greater than 0.05 (the default criterion for remaining in the model), the effect is removed.

The new model, labeled step 1 in Figure 1-10, has all three-way interactions as its generating class.

Figure 1-9
Statistics at step 0

			Effects	Chi-Square[1]	df	Sig.	Number of Iterations
Step[2]	0	Generating Class[3]	happy2*marital3*income3* sex2	.000	0	.	
		Deleted Effect	1 happy2*marital3*income3* sex2	3.839	4	.428	3

1. For 'Deleted Effect', this is the change in the Chi-Square after the effect is deleted from the model.

2. At each step, the effect with the largest significance level for the Likelihood Ratio Change is deleted, provided the significance level is larger than .050.

3. Statistics are displayed for the best model at each step after step 0.

The only effects eligible for removal are the three-way interactions, since only hierarchical models are considered. The *happy2*marital3*sex2* interaction has the largest observed significance level for the change in the chi-square if it is removed, and it is greater than 0.05, so it is eliminated from the model at step 1. The likelihood-ratio chi-square for the resulting model is 4.08, the value shown for step 2 in Figure 1-10.

Figure 1-10
Statistics used to eliminate the second effect

				Effects	Chi-Square[1]	df	Sig.	Number of Iterations
Step[2]	1	Generating Class[3]		happy2*marital3*income3, happy2*marital3*sex2, happy2*income3*sex2, marital3*income3*sex2	3.839	4	.428	
		Deleted Effect	1	happy2*marital3*income3	3.362	4	.499	5
			2	happy2*marital3*sex2	.244	2	.885	3
			3	happy2*income3*sex2	.697	2	.706	3
			4	marital3*income3*sex2	11.484	4	.022	4
	2	Generating Class[3]		happy2*marital3*income3, happy2*income3*sex2, marital3*income3*sex2	4.083	6	.665	

1. For 'Deleted Effect', this is the change in the Chi-Square after the effect is deleted from the model.

2. At each step, the effect with the largest significance level for the Likelihood Ratio Change is deleted, provided the significance level is larger than .050.

3. Statistics are displayed for the best model at each step after step 0.

At the next two steps, third-order interactions are removed, except for the *marital3*income3*sex2* interaction.

The fourth step begins with the single third-order interaction and all second-order items in the model, as shown in Figure 1-11.

Figure 1-11
Statistics at step 4

			Effects	Chi-Square[1]	df	Sig.	Number of Iterations
Step[2]	4	Generating Class[3]	marital3*income3*sex2, happy2*sex2, happy2*marital3, happy2*income3	7.714	12	.807	
		Deleted Effect	1 marital3*income3*sex2	12.002	4	.017	5
			2 happy2*sex2	2.173	1	.140	4
			3 happy2*marital3	22.548	2	.000	4
			4 happy2*income3	12.052	2	.002	4

1. For 'Deleted Effect', this is the change in the Chi-Square after the effect is deleted from the model.

2. At each step, the effect with the largest significance level for the Likelihood Ratio Change is deleted, provided the significance level is larger than .050.

3. Statistics are displayed for the best model at each step after step 0.

Since the observed significance level for the *happy2*sex2* interaction is greater than the default value of 0.05, it is removed from the model at step 4.

This results in the model shown in Figure 1-12. Now all of the interactions have observed significance levels of less than 0.05, so the removal of effects stops. The "best" model has all main effects, the two-way interactions of happiness and income and happiness and marital status, and the three-way interaction of marital status, income, and sex.

Figure 1-12
Statistics at step 5

			Effects	Chi-Square[1]	df	Sig.	Number of Iterations
Step[2]	5	Generating Class[3]	marital3*income3*sex2, happy2*marital3, happy2*income3	9.887	13	.703	
		Deleted Effect	1 marital3*income3*sex2	12.302	4	.015	5
			2 happy2*marital3	24.039	2	.000	2
			3 happy2*income3	13.534	2	.001	2

1. For 'Deleted Effect', this is the change in the Chi-Square after the effect is deleted from the model.

2. At each step, the effect with the largest significance level for the Likelihood Ratio Change is deleted, provided the significance level is larger than .050.

3. Statistics are displayed for the best model at each step after step 0.

The final model, step 6 in the pivot table output, contains the same second-order interactions and the same third-order interaction suggested by the partial-association table shown in Figure 1-8. The observed significance level for the likelihood-ratio chi-square is 0.70, suggesting that the model may be adequate. Remember that the selected model is a suggested model; there are other models that may fit the data equally well or better and may be easier to interpret. Spend some time looking for them, using the suggested model as a starting point.

You should also examine the residuals to see if any anomalies are apparent. To do this, you must rerun the analysis, specifying the final model. From the menus choose:

Analyze
  Loglinear
    Model Selection...

▶ Factor(s): happy2(1,2), marital3(1,3), income3(1,3), sex2(1,2)

Define Range for each variable (select variable and click Define Range)

Model Building
⊙ Enter in single step

Model...
  Specify Model
  ⊙ Custom
    ▶ marital3*income3*sex2
    ▶ happy2*marital3
    ▶ happy2*income3

Options...
  Plot
  ☑ Residuals
  ☑ Normal probability

You'll see that all of the standardized residuals are small in absolute value. The plots raise some concerns, since although the observed and expected values are reasonably close in value, the residuals don't appear to be randomly distributed. Examine the table to try to identify groups of cells where the model underpredicts or overpredicts, and modify the model as necessary.

Figure 1-13
Plots of residuals and cell counts

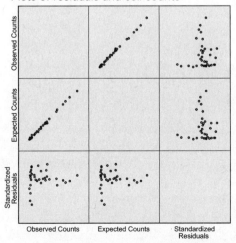

Logit Loglinear Analysis

You can use the general loglinear model (described in the *SPSS 14.0 Statistical Procedures Companion* [Norušis, 2005]) and hierarchical model selection (Chapter 1 in this book) to examine relationships among categorical variables. All variables that define the cells of the multiway crosstabulations are factors, or independent variables. The dependent variable is the number of cases in the various cells of the table.

If you can identify one of the variables as a dependent variable, you can analyze your crosstabulation with a closely related model. For example, you want to know if education level and gender are related to job satisfaction. Or, you want to know if a person's willingness to buy your product depends on age, income, and place of residence. In the first example, job satisfaction is the dependent variable, since you want to know if the other variables change the likelihood that a person is satisfied with his or her job. Similarly, in the second example, willingness to buy the product is the dependent variable.

A special class of loglinear models, called **logit models**, is used to model the relationship between one or more dependent categorical variables and a set of independent categorical variables (as well as covariates). Sometimes they are called **multinomial logit models**, since, for each combination of values of the independent variable, you assume that there is a multinomial distribution of the dependent variable and that the cell counts across combinations are independent.

Dichotomous Logit Model

Consider first an example in which the dependent variable is dichotomous—that is, it has only two categories. In the chapter "General Loglinear Analysis" in the *SPSS 14.0 Statistical Procedures Companion* (Norušis, 2005), you looked at the relationship of education, voting, and gender when none of the variables was singled out as dependent. Let's see how the analysis changes if you consider voting to be a dependent variable. That is, you want to identify factors that are associated with the likelihood that a person votes. The data are from the General Social Survey (Davis and Smith, 1972–2002).

Loglinear Representation

Figure 2-1 is a crosstabulation of voting in the last election and having a college degree.

Figure 2-1
Crosstabulation of vote and degree

Count

Vote	College Degree		Total
	No	Yes	
No	369	50	419
Yes	659	372	1031
Total	1028	422	1450

If you fit a loglinear model to the data, you have an equation that relates the log of the expected frequencies in each cell to a set of parameters. For example, the saturated model for the first cell, the count of people without degrees who do not vote, is

$$\log(m_{11}) = \mu + \lambda^{\text{nonvoter}} + \lambda^{\text{withoutdegree}} + \lambda^{\text{nonvoterwithoutdegree}}$$

Similarly, the model for people without degrees who do vote is

$$\log(m_{21}) = \mu + \lambda^{\text{voter}} + \lambda^{\text{withoutdegree}} + \lambda^{\text{voterwithoutdegree}}$$

Logit Model

If one of the variables is a dependent variable, instead of modeling the counts of cases for each cell, you model the ratio of the counts of the two values of the dependent variable for each of the combinations of values of the independent variables. For example, if education is the independent variable and voting in the last election is the dependent variable, you model the ratio of voters to nonvoters for each category of education. So, instead of having two separate models for each category of education, you have one.

For those without a degree, the log of the ratio of nonvoters to voters can be written as

$$\log\left(\frac{m_{11}}{m_{21}}\right) = \log(m_{11}) - \log(m_{21})$$

since the log of a ratio is the log of the numerator minus the log of the denominator. If you substitute the previously given equations for log(m_{11}) and log(m_{21}), you get

$$\log\left(\frac{m_{11}}{m_{21}}\right) = \lambda^{nonvoter} + \lambda^{nonvoterwithoutdegree} - \lambda^{voter} - \lambda^{voterwithoutdegree}$$

The intercept μ and $\lambda^{withoutdegree}$ don't appear in the equation, since they cancel out. The equation simplifies even further because coefficients that involve the voter category are 0, since it is the reference category.

The quantity on the left side of the equation is the log of the ratio of the count of nonvoters to voters. It is the log of the odds that a person without a degree will not vote. The log of the odds is called a **logit**, which is why these models are called logit models.

From the data shown in Figure 2-1, the observed odds that a nongraduate doesn't vote are 369/659 (0.560). It's the ratio of the number of nonvoters to the number of voters for people who don't have a degree. The log of the odds is –0.58. The observed odds that a college graduate doesn't vote are 50/372 (0.134). The log of the odds is –2.01.

The logit model always has fewer parameters than the corresponding loglinear model because the constant and all of the parameters that involve only the independent variables cancel. Every logit model with two values for the dependent variable can be converted to a loglinear model by including the appropriate terms that involve the independent variables. The parameter estimates for the logit model are the same as the parameter estimates for corresponding terms in the loglinear representation of the model. The parameter estimates from a saturated loglinear model for the data shown in Figure 2-1 are identical to those from the corresponding logit model. The goodness-of-fit statistics are the same as well.

Specifying the Model

To fit a saturated logit model, open the file *logit.sav* and from the menus choose:

Analyze
  Loglinear
    Logit...

▶ Dependent: vote
▶ Factor(s): college

Options...
  Display
    ☑ Estimates
  Criteria
    Delta: 0

A saturated logit model fits the data perfectly, so there's no need to examine any goodness-of-fit statistics. The parameter estimates from the logit model are in Figure 2-2. (The delta criterion is set to 0 so that the observed counts aren't changed by adding the default 0.5 to each cell. That's done for computational stability.)

Parameter Estimates for the Saturated Logit Model

As expected, you see only the parameter estimates for effects involving the dependent variable, *vote*. Of these, only the estimates for $\lambda^{nonvoter}$ and $\lambda^{nonvoterwithoutdegree}$ are not 0. (Remember, parameter estimates involving the last level of each factor are set to 0, so the parameters would have unique estimates.) In addition to the parameter estimates, you see that there are constants for college graduates and for those who did not graduate from college. For each combination of values of the independent variables, the constants are the sum of the parameter estimates for terms that canceled when you went from the loglinear model to the corresponding logit model.

Figure 2-2

Parameter estimates for the saturated logit model

Parameter Estimates[3,4]

Parameter		Estimate	Std. Error	Z	Sig.	95% Confidence Interval Lower Bound	95% Confidence Interval Upper Bound
Constant	[college = 0]	6.491[1]					
	[college = 1]	5.919[1]					
[vote = 0]		-2.007	.151	-13.324	.000	-2.302	-1.712
[vote = 1]		0[2]	.	.	.	.	.
[vote = 0] * [college = 0]		1.427	.164	8.698	.000	1.105	1.748
[vote = 0] * [college = 1]		0[2]	.	.	.	.	.
[vote = 1] * [college = 0]		0[2]	.	.	.	.	.
[vote = 1] * [college = 1]		0[2]	.	.	.	.	.

[1.] Constants are not parameters under the multinomial assumption. Therefore, their standard errors are not calculated.

[2.] This parameter is set to zero because it is redundant.

[3.] Model: Multinomial Logit

[4.] Design: Constant + vote + vote * college

The parameter estimates for the logit model are interpreted in the same way as the parameter estimates for the loglinear model described in the chapter "General Loglinear Analysis" in the *SPSS 14.0 Statistical Procedures Companion* (Norušis, 2005). The parameter of interest is $\lambda^{\text{nonvoterwithoutdegree}}$, which tells you whether there is a relationship between voting and degree. Note that the interaction parameter estimate shown in Figure 2-2 is identical to that in the second figure of the chapter "General Loglinear Analysis" in the *SPSS 14.0 Statistical Procedures Companion.*

Since the model is saturated, you can calculate the parameter estimate directly from the observed values. First find the log of the ratio of two odds—the odds that a person without a college degree doesn't vote and the odds that a college graduate doesn't vote. The odds that a nongraduate doesn't vote are 0.560. The odds that a college graduate doesn't vote are 0.134. The ratio of these two odds, the **odds ratio**, is 4.17. This tells you that the odds of not voting for a person without a degree is about four times the odds of not voting for a person with a degree.

The log of the odds ratio, called the **log odds ratio**, is 1.43, the value for the no college and no voting interaction. The log odds ratio tells you the difference in the log odds of a person without a degree not voting and the log odds of a graduate voting.

From Figure 2-2, the asymptotic 95% confidence interval for the log odds ratio is from 1.1 to 1.7). This translates to a 95% confidence interval for the odds ratio of 3.0 to 5.5. (If college degree and voting are independent, the expected value of the odds ratio is 1, corresponding to a log odds ratio of 0.)

Unsaturated Logit Model

In the previous example, you fit a saturated logit model with only one independent variable. You saw that college degree is related to voting. Of course, degree is only one of many possible predictors. Variables such as age may also be related to the response. Let's add one additional variable, *Age categories*, to the model. The variable has four categories: *18-29*, *30-39*, *40-49*, and *50+*. The saturated logit model includes parameters for voting, the voting-by-college interaction, the voting-by-age interaction, and the voting-by-college-by-age interaction. You will fit a logit model that omits the three-way interaction among voting, degree, and age.

Specifying the Analysis

When you specify a logit model in the Model dialog box, you specify only the factors that will have interactions with the dependent variable. SPSS creates the necessary interaction terms. To choose a model that contains all two-way interactions between the dependent variable and the independent variables, from the menus choose:

Analyze
  Loglinear
    Logit...

▶ Dependent: vote
▶ Factor(s): college agecat4

Model...
college agecat4
    Options...
      Display
        ☑ Estimates
        ☑ Frequencies
        ☑ Residuals

    Plots
        ☑ Adjusted residuals
        ☑ Normal probability for adjusted

Goodness-of-Fit Statistics

First look at Figure 2-3, which shows the goodness-of-fit statistics for the logit model with only two-way interactions. As in all statistical model building, you have to look at both the overall fit of the model and the fit of the model for individual cases.

Figure 2-3
Goodness-of-fit statistics

Goodness-of-Fit Tests [1,2]

	Value	df	Sig.
Likelihood Ratio	5.536	3	.137
Pearson Chi-Square	4.734	3	.192

[1]. Model: Multinomial Logit

[2]. Design: Constant + vote + vote * agecat4 + vote * college

The observed significance level of 0.14 for the likelihood-ratio chi-square statistics indicates that the model fits, although not as well as you would like. The plots of the residuals (not shown) do not suggest any glaring deficiencies, although the fit is somewhat poorer for older cases. For chi-square-based goodness-of-fit tests, a good model has small discrepancies between observed and expected counts, resulting in small chi-square values and large observed significance levels.

Observed and Expected Cell Counts

To determine how well the model fits individual cells, you can identify outlying points on the standardized plots of residuals, or you can examine the residuals in each of the cells. Figure 2-4 shows the observed and expected cell counts for the logit model with all two-way interactions.

Note the large standardized residuals for people in the 50+ age group. The model doesn't fit well for these cells. It may be that lumping all people over the age of 50 into a single group was a bad decision. You may want to rerun the analysis using different age groupings.

Figure 2-4
Observed and expected counts

Cell Counts and Residuals[1,2]

College Degree	Age Categories	Vote	Observed		Expected		Residual	Standardized Residual	Adjusted Residual	Deviance
			Count	%	Count	%				
No	18-29	No	107	54.3%	108.584	55.1%	-1.584	-.227	-.596	-1.773
		Yes	90	45.7%	88.416	44.9%	1.584	.227	.596	1.788
	30-39	No	84	38.5%	85.777	39.3%	-1.777	-.246	-.651	-1.875
		Yes	134	61.5%	132.223	60.7%	1.777	.246	.651	1.891
	40-49	No	60	31.4%	61.983	32.5%	-1.983	-.306	-.817	-1.975
		Yes	131	68.6%	129.017	67.5%	1.983	.306	.817	1.999
	50+	No	118	28.0%	112.656	26.7%	5.344	.588	2.163	3.307
		Yes	304	72.0%	309.344	73.3%	-5.344	-.588	-2.163	-3.255
Yes	18-29	No	17	23.6%	15.416	21.4%	1.584	.455	.596	1.823
		Yes	55	76.4%	56.584	78.6%	-1.584	-.455	-.596	-1.767
	30-39	No	17	14.0%	15.223	12.6%	1.777	.487	.651	1.937
		Yes	104	86.0%	105.777	87.4%	-1.777	-.487	-.651	-1.877
	40-49	No	12	11.5%	10.017	9.6%	1.983	.659	.817	2.082
		Yes	92	88.5%	93.983	90.4%	-1.983	-.659	-.817	-1.981
	50+	No	4	3.2%	9.344	7.5%	-5.344	-1.817	-2.163	-2.605
		Yes	121	96.8%	115.656	92.5%	5.344	1.817	2.163	3.306

[1.] Model: Multinomial Logit

[2.] Design: Constant + vote + vote * agecat4 + vote * college

For each combination of values of the independent variables, the percentages sum to 100 over the categories of the dependent variable. In Figure 2-4, you see that 54.3% of people without a college degree in the age group 18-29 did not vote. The model predicts that 55.1% will not vote. Of college graduates 50 and older, 97% voted, compared to a predicted value of 93%.

Parameter Estimates

The parameter estimates for the unsaturated model are shown in Figure 2-5. Since this is a logit model, except for the constants, only parameters that involve the dependent variable (*vote*) are displayed. In Figure 2-5, only non-zero estimates are shown. (This is not an option in SPSS; you have to edit the pivot table of estimates to remove redundant parameter estimates.)

Figure 2-5
Parameter estimates for the unsaturated model

Parameter Estimates[3,4]

Parameter	Estimate	Std. Error	Z	Sig.
[vote = 0]	-2.516	.184	-13.644	.000
[vote = 0] * [agecat4 = 1]	1.216	.166	7.303	.000
[vote = 0] * [agecat4 = 2]	.577	.163	3.540	.000
[vote = 0] * [agecat4 = 3]	.277	.175	1.579	.114
[vote = 0] * [college = 0]	1.506	.168	8.980	.000

[3.] Model: Multinomial Logit

[4.] Design: Constant + vote + vote * agecat4 + vote * college

All of the parameter estimates have log odds interpretations.

- The age-by-vote parameter estimates are the predicted log odds for not voting for each of the age groups compared to the last. For example, the log odds for not voting for a person aged 18–29 are estimated to be 1.22. That translates to an odds ratio of 3.4. That means that a person aged 18–29 is three-and-a-half times more likely to vote than not to vote than a person aged 50 or older.

- The vote-by-college parameter estimate is the predicted log odds for not voting for a person without a college degree compared to the log odds for not voting for a person with a college degree. A person without a college degree is estimated to be 4.51 times less likely to vote than not vote compared to a person with a college degree.

- To calculate predicted log odds ratios for any two categories of the independent variable, you can find the difference between the two corresponding parameter estimates. For example, the predicted log odds ratio for nonvoting for people aged 18–29 compared to people aged 30–39, is 1.22 − 0.58 = 0.64. Converted to a predicted odds ratio of 1.9, you estimate that a person aged 20–29 is almost 1.9 times more likely not to vote than to vote compared to a person aged 30–39.

Measures of Dispersion and Association

When you build a logit model, you can analyze the **dispersion**, or spread, in the dependent variable. Two statistics that are used to measure the spread of a nominal variable are Shannon's entropy measure:

$$H = -\Sigma p_j \log p_j$$

and Gini's concentration measure:

$$C = 1 - \Sigma p_j^2$$

where p_j is the probability of the event in cell j. Using either of these measures, you can subdivide the total dispersion of the dependent variable into the dispersion explained by the model and the residual (or unexplained) dispersion.

Measures of Dispersion

Figure 2-6 shows the analysis-of-dispersion table for the current model. If the model is correct, 2 times the entropy for the model has an asymptotic chi-square distribution, with the same degrees of freedom as the model. For the concentration measure, the ratio of the model concentration divided by its degrees of freedom to the residual concentration divided by its degrees of freedom has an F distribution with model and residual degrees of freedom.

Figure 2-6
Analysis-of-dispersion table

Analysis of Dispersion [1,2]

	Entropy	Concentration	df
Model	75.563	59.185	4
Residual	796.210	536.662	1445
Total	871.773	595.847	1449

[1] Model: Multinomial Logit

[2] Design: Constant + vote + vote * agecat4 + vote * college

Measures of Association

From the analysis-of-dispersion table, you can calculate statistics similar to R^2 in regression. They indicate the proportion of the total dispersion in the dependent variable that is attributable to the model (Magidson, 1981). In Figure 2-7, when dispersion is measured by the entropy criterion, the ratio of the dispersion explained by the model to the total dispersion is 0.087. When measured by the concentration criterion, it is 0.099. These values can be interpreted as measures of association. For a two-way table, the concentration measure is the square of Kendall's tau-*b*, while the entropy measure is the same as the uncertainty coefficient with response as the dependent variable. Although it is tempting to interpret the magnitude of these measures similarly to R^2 in regression, this may be misleading, since the coefficients may be small even when the variables are strongly related (Haberman, 1982). The coefficients are best interpreted in light of experience.

Figure 2-7
Measures of association

Measure of Association[1,2]

Entropy	.087
Concentration	.099

[1] Model: Multinomial Logit

[2] Design: Constant + vote + vote * agecat4 + vote * college

Polychotomous Logit Model

In the previous examples, the dependent variable had two categories, which is the simplest example of a logit model. Logit models can also be used for categorical dependent variables that have more than two values. Such models are called **polychotomous** (or **polytomous**) **logit models**. As an example, consider the relationship between marital status, age, and gender. You want to know if the odds of being in different marital status categories change with age and with gender. *Marital status*, the dependent variable, has four categories: *Never married*, *Widowed*, *Divorced/Separated* and *Married*. The age categories are *18–29*, *30–45*, *46-65*, and *66+*. The data are from the General Social Survey.

When you have a dichotomous dependent variable, you model the ratio of the two possible outcomes in each cell of the table. The only choice available to you is which outcome is the reference category. When the dependent variable has more than two values, you can construct many odds ratios for the same combination of values of the independent variables. For example, for each combination of age and gender, you can look at the odds of being *Never married* compared to *Married*, or you can look at the odds of being *Widowed* compared to *Divorced/Separated*. If you use a single category as the reference category, the model is called a **baseline category logit model**.

The Logit procedure, just like the General Loglinear procedure, considers the last category of each variable the reference category. For this example, currently married and female are the reference categories.

Specifying the Model

The logit model includes only main effects for age and female, corresponding to two-way interactions between marital status and age and marital status and gender. The three-way interaction, which allows the relationship between marital status and a variable to differ depending on the values of the third variable, is excluded. To fit a polychotomous logistic regression, from the menus choose:

Analyze
  Loglinear
    Logit...

▶ Dependent: maritalstatus
▶ Factor(s): agecat female

Model...
female agecat

Options...

Display
☑ Frequencies
☑ Residuals
☑ Estimates

Plot
☑ Adjusted residuals
☑ Normal probability for adjusted

Goodness of Fit of the Model

The goodness-of-fit statistics for fitting a polychotomous logit model without the three-way interaction of gender, age, and marital status are shown in Figure 2-8.

Figure 2-8
Goodness-of-fit statistics

Goodness-of-Fit Tests [1,2]

	Value	df	Sig.
Likelihood Ratio	8.558	9	.479
Pearson Chi-Square	9.018	9	.436

[1.] Model: Multinomial Logit

[2.] Design: Constant + maritalstatus + maritalstatus * agecat + maritalstatus * female

The observed significance level is reasonably large, indicating that, overall, the model appears to fit the data. You must examine residual plots to look for cells that the model may not fit well.

Interpreting Parameter Estimates

To see how the independent variables relate to the dependent variable, look at the parameter estimates in Figure 2-9. For readability, parameters that are set to 0 are excluded from the table. That means all parameter estimates that involve *agecat* = 4 (66+) and *female* = 1 (female) are excluded.

Figure 2-9

Parameter estimates for non-zero parameters for polychotomous model

Parameter Estimates[3,4]

Parameter	Estimate	Std. Error	Z	Sig.
[maritalstatus = 1]	-2.673	.358	-7.472	.000
[maritalstatus = 2]	.633	.157	4.030	.000
[maritalstatus = 3]	-1.427	.246	-5.792	.000
[maritalstatus = 1] * [agecat = 1]	3.079	.370	8.327	.000
[maritalstatus = 1] * [agecat = 2]	1.262	.366	3.450	.001
[maritalstatus = 1] * [agecat = 3]	.373	.401	.929	.353
[maritalstatus = 2] * [agecat = 1]	-20.963	1950.284	-.011	.991
[maritalstatus = 2] * [agecat = 2]	-5.271	.724	-7.282	.000
[maritalstatus = 2] * [agecat = 3]	-2.023	.227	-8.923	.000
[maritalstatus = 3] * [agecat = 1]	.103	.339	.304	.761
[maritalstatus = 3] * [agecat = 2]	.686	.261	2.630	.009
[maritalstatus = 3] * [agecat = 3]	.640	.269	2.376	.017
[maritalstatus = 1] * [female = 0]	.191	.155	1.231	.218
[maritalstatus = 2] * [female = 0]	-1.581	.238	-6.628	.000
[maritalstatus = 3] * [female = 0]	-.620	.152	-4.084	.000

[3.] Model: Multinomial Logit

[4.] Design: Constant + maritalstatus + maritalstatus * agecat + maritalstatus * female

To make it easier to interpret, the parameter estimates are displayed in Table 2-1. The table contains the parameter estimate λ and in parentheses e^{λ}. Note that there is a problem with estimating the parameter for marital status widowed for the age group 18–29. That's because there are no young widows in the data file.

Table 2-1
Parameter estimate (e^{λ}) summary

	Marital Status			
Age	**Never married**	**Widowed**	**Divorced/Separated**	**Married**
18–29	3.08 (21.76)	−20.96 (< 0.001)	0.10 (1.11)	0 (1)
30–45	1.26 (3.52)	−5.27 (0.005)	0.69 (1.99)	0 (1)
46–65	0.37 (1.45)	−2.02 (0.130)	0.64 (1.90)	0 (1)
66+	0 (1)	0 (1)	0 (1)	0 (1)
Gender				
Male	0.19 (1.21)	−1.58 (0.200)	−0.62 (0.54)	0 (1)
Female	0 (1)	0 (1)	0 (1)	0 (1)

Parameter Estimates for Age

The parameter estimate for age group 18–29 and never married is 3.08. The value of e^{λ} is 21.76. This tells you that based on the model, you are almost 22 times more likely to be never married than married at age 18–29, compared to being never married to married at age 66+. In Table 2-1, you see that the likelihood of being never married to married decreases with age. However, in Figure 2-9, you see that the parameter estimate for age group 46–65 is not significantly different from 0. This means that by this age group, the odds of being never married to married are not significantly different from those aged 66+.

Now look at the widowed row. The parameter estimate for age group 18–29 is very large and negative. That's because the odds of being widowed to being married at age 18–29 are very small compared to those at age 66+. Notice in Figure 2-9 that the standard error of the parameter estimate is quite large and that the parameter estimate is not significantly different from 0. This is because there are no young widowed cases in the sample, so the parameter cannot be estimated well. The correct conclusion is not that the value of the parameter is close to 0 but that it can't be estimated well from the data.

For age group 30–45 compared to age group 66+, the estimated odds ratio of being widowed to married is 0.005. This means that the odds of being widowed to married are almost 200 times as large for people at age 66+ when compared to age group 30–45. Notice that this parameter estimate has a much smaller standard error than the previous parameter estimate and *is* significantly different from 0.

Parameter Estimates for Gender

Now consider how a person's gender relates to marital status. The odds ratio of 1.2 for never-married males tells you that for a male, the estimated odds of being never married to married are 1.2 times as large as the same odds for a female. (Since the variable in the model is coded 1 for females, the parameter estimate shown is for males.) In Figure 2-9, you see that the parameter estimate is not significantly different from 0, so you can't conclude that the observed odds ratio of 1.2 is significantly different from 1, the odds ratio under the null hypothesis.

The estimated odds ratio for widowed males is 0.20. This means that the odds for a male of being widowed to married are one-fifth as large as the same odds for a female. Similarly, the estimated odds ratio for divorced or separated males tells you that the odds that a male is divorced or separated compared to married are a little more than half of the odds that a female is divorced or separated compared to married. Thus, at any age, a female is almost two times as likely as a male to be divorced or separated compared to married.

Predicted Odds for Cells

From the parameter estimates, for each cell in the two-way table of age and gender, you can calculate a predicted odds for any value of the dependent variable compared to the reference category. For example, the predicted odds for being a never-married (marital category 1) male at age 18–29 (age category 1), compared to being a married male at age 18–29 is

$$e^{\lambda^{\text{marital1}} + \lambda^{\text{agecat1marital1}} + \lambda^{\text{malemarital1}}} = 1.81$$

where from Figure 2-9, the parameter estimates are for *marital status 1* = –2.673; *agecat 1*-by-*marital status 1* = 3.079; *male*-by-*marital 1*= 0.19. This is the ratio of the predicted number of never-married males of age 18–29 to the predicted number of married males of age 18–29.

Note: In Figure 2-9, *male-by-marital 1* is shown as *(maritalstatus=1)(female=0)*.

Similarly, for females, the predicted odds of being never married at age 46-65, compared to being married at age 46–65, are only 10%.

$$e^{\lambda^{marital1} + \lambda^{agecat3marital1} + \lambda^{femalemarital1}} = 0.10$$

where from Figure 2-9, the parameter estimates are for *maritalstatus category 1* = –2.673; *agecat 3-by-maritalstatus category 1* = 0.373; *female-by-maritalstatus category 1* = 0 (since all parameter estimates for females are 0). This means that a woman at age 46–65 is 10 times more likely to be currently married than never married.

Observed and Predicted Cell Counts

Figure 2-10 shows part of the table of observed and expected cell counts and residuals for the current analysis. The percentages sum to 100 for each combination of values of the independent variables. You see that the expected number of never-married males at age 18–29 is 74.86. The number of married males is 41.24. The ratio of the expected counts of never married to married is 1.81, the odds you calculated previously from the parameter estimates.

For women aged 46–65, the expected number of never-married women is 12.88, and the number of married women is 128.58. The odds of being never married to currently married are 10%, the value previously calculated.

Figure 2-10
Observed and expected cell counts and residuals

Cell Counts and Residuals[1,2]

Age category	female	Marital status	Observed Count	%	Expected Count	%
18-29	No	Never married	76	62.3%	74.860	61.4%
		Widowed	0	.0%	.000	.0%
		Divorced/Separated	6	4.9%	5.899	4.8%
		Married	40	32.8%	41.241	33.8%
46-65	Yes	Never married	9	3.9%	12.883	5.6%
		Widowed	31	13.4%	32.030	13.8%
		Divorced/Separated	60	25.9%	58.512	25.2%
		Married	132	56.9%	128.575	55.4%

[1.] Model: Multinomial Logit

[2.] Design: Constant + maritalstatus + maritalstatus * agecat + maritalstatus * female

Examining Residuals

The same diagnostic residual plots used for loglinear models are also used for the logit model. Figure 2-11 is a scatterplot matrix of the adjusted residuals. No problems are readily apparent. Figure 2-12, the normal probability plot, appears consistent with normality.

Figure 2-11
Scatterplot matrix of adjusted residuals

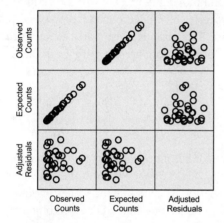

Figure 2-12
Normal probability plot of adjusted residuals

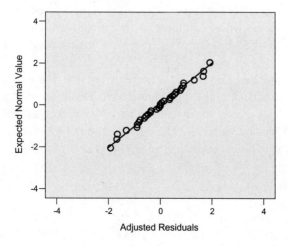

Covariates

You can incorporate covariates into logit models, just like you do in the general loglinear model. The average value for the covariate is used for all observations in the cell. Consider the job satisfaction and earnings example in the chapter "General Loglinear Analysis" in the *SPSS 14.0 Statistical Procedures Companion* (Norušis, 2005). If job satisfaction is considered the dependent variable, you can specify the row effects model in the Logit procedure by moving *jobsatisfact* into the Dependent list, moving *earncat* into the Factor list, and moving *cov* (a copy of the earning category variable) into the Covariate list. To specify the model, click Model, select Custom and move *cov* into the Terms in Model list. The parameter estimates of interest and the goodness-of-fit statistics from the Logit procedure are the same from the loglinear model.

Other Logit Models

In the previous example, the reference (or baseline) category for comparisons was one of the values of the dependent variable. (SPSS always chooses the last category of the dependent variable as the baseline category. By recoding the values of your dependent variable, you can select any category to be the baseline.) There are other ways to choose the comparisons. For example, when the dependent variable is ordinal, you can fit an **adjacent-categories logit model**. In an adjacent-categories logit model, each category of the dependent variable is compared to the next category. In a **continuation-ratio logit model**, each category of the dependent variable is compared to the sum of the categories that follow. SPSS will not fit these models directly; however, it is possible to fit the models in SPSS by estimating a series of logit models. See Agresti (2002) for details.

The models in this chapter include a single dependent variable. However, you can specify more than one dependent variable in a logit model. The interaction terms let you examine the relationships among the dependent variables as well as their interactions with the independent variables. The interpretation of these interaction terms is the same as for the identical terms in a general loglinear model.

Multinomial Logistic Regression

When you have a dependent variable that is **dichotomous**—that is, it can have only two values—you can use binary logistic regression to model the relationship between the dependent variable and a set of independent or predictor variables. For example, you can model the probability that someone will purchase your product based on characteristics such as age, education, gender, and income. For more information about two-group logistic regression, see the *SPSS 14.0 Statistical Procedures Companion* (Norušis, 2005).

If you have a categorical dependent variable with more than two possible values, you can use an extension of the binary logistic regression model—called **multinomial** (or **polytomous**) **logistic regression**—to examine the relationship between the dependent variable and a set of predictor variables. The models are called multinomial because, for each combination of values (or covariate pattern) of the independent variables, the counts of the dependent variable are assumed to have a multinomial distribution. The counts at the different combinations are also assumed to be independent with a fixed total. If your dependent variable is ordinal (such as severe, moderate, and minimal), you can fit ordinal logistic regression models, such as those described in Chapter 4 (see Agresti, 2002).

SPSS has two main procedures that are designed for estimating logistic regression models: the Binary Logistic Regression procedure and the Multinomial Logistic Regression procedure. Both procedures can be used for binary regression models. The Multinomial Logistic Regression procedure aggregates the data, while the Binary Logistic Regression procedure does not. This affects goodness-of-fit testing.

The SPSS Multinomial Logistic Regression procedure, described in this chapter, can also be used to analyze data from one-on-one matched case-control studies. These are studies that construct matched pairs of cases—that is, one has experienced the event of interest and the other has not. Case-control studies are often used in medicine to identify predictors of the event. For more information, see "Matched Case-Control Studies" on p. 63.

The Logit Model

When you have two groups in which one has experienced an event of interest and the other has not, you can write the logistic regression model as

$$\log\left(\frac{P(\text{event})}{P(\text{noevent})}\right) = B_0 + B_1X_1 + B_2X_2 + \dots + B_pX_p$$

where B_0 is the intercept, B_1 to B_p are the logistic regression coefficients, and X_1 to X_p are the independent variables. The quantity on the left side of the equals sign is called a **logit**. It is the natural log of the odds that the event occurs. When the dependent variable has only two values, there is only one unique logit. This is because switching which of the two values is the "event" results in the same logistic regression coefficients for the independent variables with the signs reversed.

If your dependent variable has *J* possible values, the number of nonredundant logits you can form is $J - 1$. The simplest type of logit for this situation is called the **baseline category logit**. For each group, you calculate the log of the ratio of the probability of being in that group compared to being in the baseline group. If the baseline category is *J*, for the ith category, the model is

$$\log\left(\frac{P(\text{category}_i)}{P(\text{category}_j)}\right) = B_{i0} + B_{i1}X_1 + B_{i2}X_2 + \dots + B_{ip}X_p$$

You have a set of coefficients for each logit, which is why each coefficient in the previous equation has two subscripts: the first identifies the logit and the second identifies the variable. For the baseline category, the coefficients are all 0. For example, if the dependent variable has three values, you will generate two sets of non-zero coefficients—one for the comparison of each of the first two groups to the last group.

Baseline Logit Example

As an example of multinomial logistic regression, you'll consider three groups of people: those who feel the U.S. spends too little on space exploration, those who feel the amount is just right, and those who feel the amount is too much. The data are from the General Social Survey (Davis and Smith, 1972–2002). Since the categories are ordered from too little to too much, you can also analyze these data using the SPSS Ordinal Regression procedure, described in Chapter 4. The Ordinal Regression procedure assumes that the relationship between the logits and the variables is the same for each logit. Except for the intercept, all logits have the same coefficients for the independent variables.

Let's start simply by considering only one independent variable: gender. Do males and females perceive spending on space exploration differently? (Of course, you can handle the single-variable problem by computing a chi-square test of independence, but it's easier to see what's going on in multinomial regression by first looking at equations with a single independent variable. Adding more variables to the equations is then easy.)

Figure 3-1 is a crosstabulation of the responses by gender. From the row percentages, you see that 16% of males and 8.5% of females think that too little money is being spent on space exploration; 43.4% of females and only 33% of males feel that too much money is being spent. Approximately equal percentages feel that the amount is just right.

Figure 3-1
Crosstabulation of responses by gender

			Resources on space program			
			Too little	Just right	Too much	Total
Gender	Female	Count	121	680	615	1416
		% within Gender	8.5%	48.0%	43.4%	100.0%
	Male	Count	192	607	385	1184
		% within Gender	16.2%	51.3%	32.5%	100.0%
Total		Count	313	1287	1000	2600
		% within Gender	12.0%	49.5%	38.5%	100.0%

Since the dependent variable has three categories, you can form two nonredundant logits. If you use *Just right* as the base, or reference, category and *gender* as the single independent variable:

$$g_1 = \log\left(\frac{P(\text{Toolittle})}{P(\text{Justright})}\right) = B_{10} + B_{11}(\text{female})$$

$$g_2 = \log\left(\frac{P(\text{Toomuch})}{P(\text{Justright})}\right) = B_{20} + B_{21}(\text{female})$$

The SPSS Multinomial Logistic Regression procedure treats the last category of a categorical or factor variable as the reference category and sets its coefficient to 0. The variable *male* is coded 0 for *no* and 1 for *yes*, so males are the reference category.

Specifying the Model

To fit the logistic regression model with three groups and one independent variable, open the file *space.sav*, and from the menus choose:

Analyze
  Regression
    Multinomial Logistic...

▶ Dependent: space

Reference Category...
  ⊙ Custom
    Value: 2

▶ Factor(s): male

Statistics...
  Parameters
    ☑ Estimates

The reference category for the logits is *Just right*, coded as 2. The default reference category is the last.

Parameter Estimates

The two sets of logistic regression coefficients are shown in Figure 3-2. Using these coefficients, the estimated logit equations for females are:

$$g_1 = \log\left(\frac{P(Toolittle)}{P(Justright)}\right) = -1.15 - 0.575$$

$$g_2 = \log\left(\frac{P(Toomuch)}{P(Justright)}\right) = -0.455 + 0.355$$

Figure 3-2

Parameter estimates for model with intercept and gender

Resources on space program [1]		B	Std. Error	Wald	df	Sig.	Exp(B)	95% Confidence Interval for Exp(B)	
								Lower Bound	Upper Bound
Too little	Intercept	-1.151	.083	193.250	1	.000			
	[male=0]	-.575	.129	19.947	1	.000	.563	.437	.724
	[male=1]	0^2	.	.	0	.	.	.	.
Too much	Intercept	-.455	.065	48.832	1	.000			
	[male=0]	.355	.086	17.148	1	.000	1.426	1.205	1.687
	[male=1]	0^2	.	.	0	.	.	.	.

[1]. The reference category is: Just right.

[2]. This parameter is set to zero because it is redundant.

For males, the estimated value of each logit is just the intercept, since the coefficient for the predictor variable is 0. The log of the ratio of the probability of a male choosing *Too little* to the probability of a male choosing *Just right* is –1.15, the intercept for the first logit. (You can calculate the value from Figure 3-1 by taking the natural log of the ratio of males who choose *Too little* [192] to the number of males who choose *Just right* [607].) The value of the second logit for males is the log of the ratio of the probability of a male choosing *Too much* to the probability of a male choosing *Just right*.

The coefficients for *male* tell you about the relationship between the logits and *gender*. For the first logit, the coefficient is negative and significantly different from 0, indicating that females are less likely than males to select *Too little*. From Figure 3-1, the estimated odds for a female selecting *Too little* compared to *Just right* are 121/680. For a male, the estimated odds for *Too little* compared to *Just right* are 192/607. The ratio of these two odds (the odds ratio, 0.563) is in the column labeled *Exp(B)* in Figure 3-2. A male is 1.78 times as likely as a female to choose *Too little* compared to *Just right*. (The odds ratio for comparing males to females is the reciprocal of the odds ratio that compares females to males.)

The coefficient for the second logit is positive, indicating that compared to males, females are more likely to select *Too much* rather than *Just right*. In the column labeled *Exp(B)*, you see that females are 40% more likely than males to select *Too much* rather than *Just right*.

Estimating the Redundant Logit

The coefficients in Figure 3-2 describe the relationship between *gender* and the two logits, with *Just right* as the reference category. An additional pairwise comparison that you can make is *Too little* compared to *Too much*. Since this is a redundant logit, the coefficients for this comparison are the difference of the two sets of coefficients you have already estimated. This is because:

$$\log\left(\frac{(\text{Toolittle})}{P(\text{Toomuch})}\right) = \log\left(\frac{(\text{Toolittle})}{P(\text{Justright})}\right) - \log\left(\frac{P(\text{Toomuch})}{P(\text{Justright})}\right)$$

(Remember that $\log(a/b) = \log(a) - \log(b)$.) If you are interested in this logit, you can specify *Too much* as the reference category. The procedure will then automatically calculate the coefficients and standard errors of interest. You can obtain tests for linear combinations of parameters using the TEST subcommand in syntax.

Adding Prestige to the Model

You have seen that gender appears to be related to views about spending on space exploration. Now consider whether occupational prestige is also a significant predictor when it is added to a model that contains gender. (Low scores on the prestige scale are assigned to occupations that are not "prestigious.")

Recall the Multinomial Logistic Regression dialog box and select:

▶ Covariate(s): prestige

Statistics...
    Model
    ☑ Model fitting information
    Parameters
    ☑ Estimates
    ☑ Likelihood ratio tests

Figure 3-3 contains the parameter estimates for the model with *gender* and *prestige*. In this table, you see that including *prestige* has not much changed the coefficients for *gender*. You can also see that the *prestige* parameter estimate is not significantly different from 0 for the first logit but is significantly different for the second logit. This indicates that *prestige* is not significantly related to the separation of the *Too little* and *Just right* groups but that it is useful in separating the *Too much* and *Just right* groups.

Figure 3-3
Parameter estimates for gender and prestige

Resources on space program [1]		B	Std. Error	Wald	df	Sig.	Exp(B)	95% Confidence Interval for Exp(B) Lower Bound	Upper Bound
Too little	Intercept	-1.373	.224	37.685	1	.000			
	prestige	.005	.005	.990	1	.320	1.005	.996	1.014
	[male=0]	-.578	.132	19.245	1	.000	.561	.434	.726
	[male=1]	0 [2]	.	.	0	.	.	.	.
Too much	Intercept	.345	.152	5.186	1	.023			
	prestige	-.019	.003	34.788	1	.000	.981	.975	.987
	[male=0]	.349	.088	15.700	1	.000	1.418	1.193	1.685
	[male=1]	0 [2]	.	.	0	.	.	.	.

[1]. The reference category is: Just right.

[2]. This parameter is set to zero because it is redundant.

The odds ratio for *prestige* is close to 1 for both logits. Prestige scores are measured on a scale from 10 to 90. The odds ratio tells you the change in odds for a one-unit change in occupational prestige. A one-unit change isn't meaningful on a scale like this. For continuous variables, you should look at the change associated with a meaningful interval for your data. For example, a 20-point difference might be reasonable for the prestige score, while $10,000 might be meaningful for income.

Checking Linearity

The multinomial regression model requires a linear relationship between the predictor variables and the logits. For a dichotomous variable, or categorical variables that are transformed prior to analysis, that's not a problem. For other variables, the linearity assumption must be checked. One way of checking is to create a categorical variable that corresponds to equal intervals of the *prestige* variable and enter that variable into the model instead of the continuous version. Then examine the estimated coefficients to see whether they increase or decrease in an approximately linear fashion.

Recall the Multinomial Logistic Regression dialog box and select:

▶ Factor(s): male, prestige4

Figure 3-4
Coefficients for categorical prestige variable

Resources on space program [1]		B	Std. Error	Wald	df	Sig.	Exp(B)
Too little	Intercept	-1.001	.259	14.909	1	.000	
	[male=0]	-.583	.133	19.332	1	.000	.558
	[male=1]	0^2	.	.	0	.	.
	[prestige4=1]	-.128	.284	.202	1	.653	.880
	[prestige4=2]	-.305	.275	1.225	1	.268	.737
	[prestige4=3]	.034	.291	.014	1	.906	1.035
	[prestige4=4]	0^2	.	.	0	.	.
Too much	Intercept	-.819	.215	14.448	1	.000	
	[male=0]	.354	.088	16.082	1	.000	1.425
	[male=1]	0^2	.	.	0	.	.
	[prestige4=1]	.600	.226	7.056	1	.008	1.823
	[prestige4=2]	.339	.221	2.355	1	.125	1.404
	[prestige4=3]	-.033	.237	.020	1	.888	.967
	[prestige4=4]	0^2	.	.	0	.	.

[1] The reference category is: Just right.

[2] This parameter is set to zero because it is redundant.

From the observed significance levels for the first logit, you see that *prestige* does not appear to be related to the selection of *Too little* compared to *Just right*, so there's not much reason to examine the linearity of the coefficients, since 0 is a plausible value for all of them. *Prestige* is related to the selection of *Too much* compared to *Just right*.

Figure 3-5 shows a plot of the logit coefficients for the second logit. (To obtain the plot, activate the pivot table with the coefficients by double-clicking it. Then highlight the coefficients for *prestige* in the *Too much* category and right-click the mouse. Select Create Graph, Line.) The first three coefficients are linearly related to the *prestige* category. The coefficient decreases by about 0.3 for each unit increase in the *prestige* category. However, the coefficient for the third category is very close to 0, the value for the fourth category. It appears that after a certain amount of occupational prestige, additional "prestige" doesn't matter, which suggests that it may make sense to combine the third and fourth categories of *prestige*. It also suggests that treating the relationship as linear over the range of values of the variable may be problematic. For later analysis of this data, you'll combine the last two categories of *prestige* into a single category.

Figure 3-5
Plot of logit coefficients

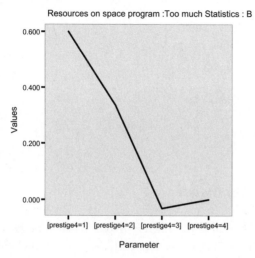

Resources on space program :Too much Statistics : B

Likelihood-Ratio Test for Individual Effects

In the parameter estimates table shown in Figure 3-4, the Wald test is used to test the null hypothesis that, for each logit, each of the individual coefficients is 0. However, tests based on the Wald statistic have undesirable properties. For example, they sometimes fail to correctly reject the null hypothesis when coefficients are large (see Hauck and Donner, 1977).

The likelihood-ratio test for each effect, shown in Figure 3-6, has better statistical properties and provides an overall test of each effect. The test is based on the change in the value of –2 log-likelihood if the effect is removed from the final model. If all coefficients for an effect are 0, this change has a chi-square distribution with degrees of freedom equal to the degrees of freedom for the effect being removed.

From Figure 3-6, you can conclude that both *gender* and *prestige* are significantly related to views on space exploration. The change in –2 log-likelihood is significant if *gender* is removed from the model containing the intercept, *gender*, and *prestige*. There is also a significant change if *prestige* is removed from the model containing the intercept, *gender*, and *prestige*.

Figure 3-6
Likelihood-ratio tests

Effect	-2 Log Likelihood of Reduced Model	Chi-Square	df	Sig.
Intercept	91.082[1]	.000	0	.
male	141.191	50.109	2	.000
prestige4	127.921	36.839	6	.000

The chi-square statistic is the difference in -2 log-likelihoods between the final model and a reduced model. The reduced model is formed by omitting an effect from the final model. The null hypothesis is that all parameters of that effect are 0.

[1] This reduced model is equivalent to the final model because omitting the effect does not increase the degrees of freedom.

Likelihood-Ratio Test for the Overall Model

The likelihood-ratio test can be used to test the null hypothesis that all coefficients in the model are 0. You compute the value of –2 log-likelihood for a model with only an intercept term and for a model with all of the effects of interest. The difference between these values is shown in the column labeled *Chi-Square*. If the observed significance level is small, you can reject the null hypothesis that all coefficients for *gender* and *prestige* are 0. You can conclude that the final model is significantly better than the intercept-only model.

Figure 3-7
Model-fitting information

Model	Model Fitting Criteria	Likelihood Ratio Tests		
	-2 Log Likelihood	Chi-Square	df	Sig.
Intercept Only	174.891			
Final	86.167	88.723	14	.000

The log-likelihood can be expressed as the sum of a multinomial constant that doesn't depend on the parameters and the kernel, a quantity that does depend on the parameters. The –2 log-likelihood values shown in Figure 3-7 include both the constant and the kernel. Many books and programs, including the SPSS Logistic Regression procedure, report only the kernel values. Since most tests are based on differences of log-likelihoods, the constants do not matter. If the number of cases is equal to the number of covariate patterns, the constant is 0.

Including an Interaction Effect

So far, you have considered the effect of *gender* and *prestige*. Both of these variables are significantly related to attitudes on money spent for space exploration. You looked at the effects of the variables individually, but you haven't considered a possible interaction between *gender* and *prestige*. There is an interaction effect between *gender* and *prestige* if the effect of prestige is not the same for men and women. For example, it's possible that men with highly prestigious jobs favor spending on space exploration even more than you would predict, based only on the coefficients for just *male* and *prestige4*.

Recall the Multinomial Logistic Regression dialog box and select:

Model...
  Specify Model
  ⊙ Full Factorial

Figure 3-8
Likelihood-ratio tests for effects in the model

Effect	Model Fitting Criteria	Likelihood Ratio Tests		
	-2 Log Likelihood of Reduced Model	Chi-Square	df	Sig.
Intercept	86.167[1]	.000	0	.
male	86.167[1]	.000	0	.
prestige4	86.167[1]	.000	0	.
male * prestige4	91.082	4.914	6	.555

The chi-square statistic is the difference in -2 log-likelihoods between the final model and a reduced model. The reduced model is formed by omitting an effect from the final model. The null hypothesis is that all parameters of that effect are 0.

[1] This reduced model is equivalent to the final model because omitting the effect does not increase the degrees of freedom.

Figure 3-8 shows that when the interaction of *gender* and *prestige* is removed from the model, the change in the −2 log-likelihood is not large enough to reject the null hypothesis that all of the coefficients associated with the interaction effect are 0. The intercept and main effects of *gender* and *prestige* are included in the interaction, and removing them doesn't change the fit of the model. The likelihood-ratio test is not calculated for these effects.

Evaluating the Model

There are several ways to evaluate how well your model fits the data. Let's return to the main-effects-only model. Recall the Multinomial Logistic Regression dialog box and select:

Model...
    Specify Model
      ⊙ Main effects

Statistics...
    Model
      ☑ Pseudo R-square
      ☑ Cell probabilities
      ☑ Classification table
      ☑ Goodness-of-fit

Calculating Predicted Probabilities and Expected Frequencies

From the logistic regression model coefficients, you can estimate the probability that a person will have each of the three opinions. As an example, to calculate the probability that a woman with the lowest level of the prestige score has a particular opinion, you must estimate the values of each of the three logits, using the values for the intercept and the coefficients for female and a prestige score of 1 (from Figure 3-4):

$$P(\text{group}_i) = \frac{\exp(g_i)}{\displaystyle\sum_{k=1}^{J} \exp(g_k)}$$

$g_1 = -1.001 - 0.583 - 0.128 = -1.712$

$g_2 = 0$

$g_3 = -0.819 + 0.354 + 0.600 = 0.135$

$$P(\text{Too little}) = \frac{0.180}{(0.180 + 1 + 1.14)} = 0.077$$

$$P(\text{Justright}) = \frac{1}{(0.180 + 1 + 1.14)} = 0.431$$

$$P(\text{Toomuch}) = \frac{1.14}{(0.180 + 1 + 1.14)} = 0.492$$

There are 375 females with the prestige category of 1 in the data set. Based on the estimated probabilities (calculated to more decimal places), you predict that 29.12 think that *Too little* is spent on space exploration, 161.26 think that the amount spent is *Just right*, and 184.62 think that *Too much* is spent.

In the table of observed and predicted frequencies shown in Figure 3-9, only the first prestige category is shown. You see that 33 females think that *Too little* is spent on space exploration, 158 think that the amount spent is *Just right*, and 184 think that *Too much* is spent. For each cell, the Pearson residual is also calculated. The standardized Pearson residual is the difference between the observed and predicted cell counts divided by an estimate of the standard deviation.

$$\text{standardized residual} = \frac{O_{ij} - E_{ij}}{\sqrt{n_i \hat{p}_{ij}(1 - \hat{p}_{ij})}}$$

The Pearson residuals are used to assess how well a model fits the observed data. Cells with Pearson residuals greater than 2 in absolute value should be examined to see if there is an identifiable reason why the model does not fit well. None of the residuals in Figure 3-9 is particularly large.

Figure 3-9
Observed and predicted frequencies and residuals for a model with gender and prestige

Occupational prestige scale	Gender	Resources on space program	Frequency			Percentage	
			Observed	Predicted	Pearson Residual	Observed	Predicted
1	Female	Too little	33	29.118	.749	8.8%	7.8%
		Just right	158	161.264	-.340	42.1%	43.0%
		Too much	184	184.618	-.064	49.1%	49.2%
	Male	Too little	51	54.882	-.569	14.1%	15.2%
		Just right	173	169.736	.344	47.9%	47.0%
		Too much	137	136.382	.067	38.0%	37.8%

The percentages are based on total observed frequencies in each subpopulation.

Classification Table

If you classify each case into the group for which it has the highest predicted probability, you can compare the observed and predicted groups, as shown in Figure 3-10.

Figure 3-10
Classification table

	Predicted			
Observed	Too little	Just right	Too much	Percent Correct
Too little	0	267	33	.0%
Just right	0	1093	158	87.4%
Too much	0	760	184	19.5%
Overall Percentage	.0%	85.0%	15.0%	51.2%

Figure 3-10 shows that the model does poorly in identifying people who chose *Too little*. None of them is classified correctly. Of the 1,251 people who chose *Just right*, 1,093 (87.4%) are correctly assigned to *Just right* by the model. Only 20% of the *Too much* category is correctly classified. Overall, about half of the respondents are correctly assigned. Does this mean that the model doesn't fit? Not necessarily. It is possible for the model to be correct and informative but for classification to be poor. When you have groups of unequal sizes, as in this example, cases will be more likely to be classified to the larger groups regardless of how well the model fits. Although a classification table provides interesting information, by itself it tells you little about how well a model fits the data. (For more information, see Hosmer and Lemeshow, 2000.)

Goodness-of-Fit Tests

Whenever you build a model, you are interested in knowing how well it fits the observed data. The Pearson chi-square statistic is often used to assess the discrepancy between observed and expected counts in a multidimensional crosstabulation. It is computed in the usual manner as

$$\chi^2_{\text{Pearson}} = \sum_{\text{all cells}} \frac{(\text{observed count} - \text{expected count})^2}{\text{expected count}}$$

Large values for the Pearson χ^2 indicate that the model does not fit well. If the observed significance level is small, you can reject the null hypothesis that your model fits the observed data. Another measure of goodness of fit is the deviance chi-square. It is the change in –2 log-likelihood when the model is compared to a *saturated* model—that is, when it is compared to a model that has all main effects and interactions. If the model fits well, the difference between the log-likelihoods should be small and the observed significance level should be large. For large sample sizes, both goodness-of-fit statistics should be similar.

The degrees of freedom for both statistics depend on the number of distinct observed combinations of the independent variables (often called the number of covariate patterns), the number of independent logits, and the number of parameters estimated. In this example, the two independent variables form 8 cells, all of them with observed counts greater than 0; the number of independent logits is 2; and the total number of estimated nonredundant parameters, including the intercept, is 10. The degrees of freedom are the product of the number of observed covariate patterns and the number of independent logits minus the number of estimated parameters: $(8 \times 2) - 10 = 6$.

Figure 3-11
Goodness-of-fit statistics for a model with gender and prestige

	Chi-Square	df	Sig.
Pearson	4.877	6	.560
Deviance	4.914	6	.555

Figure 3-11 shows that you cannot reject the null hypothesis that the model fits. Notice that the deviance chi-square in Figure 3-11 is the same as the likelihood-ratio chi-square for the interaction of *gender* and *prestige* in Figure 3-8. That's not a coincidence. The saturated model for Figure 3-11 is the model with *gender*, *prestige*, and the interaction of *gender* and *prestige*. In both Figure 3-11 and Figure 3-8, you are looking at the change in –2 log-likelihood when the interaction is removed from the saturated model.

Specifying Covariate Patterns

The goodness-of-fit statistics are based on comparing observed and expected cell counts in each cell of the table. The number of cells is the product of the number of levels of the grouping variable and all of the factors and covariates. Each distinct combination of values of the predictor variables is known as a **covariate pattern**. The goodness-of-fit statistics should be used only when there are multiple cases observed for each of the covariate patterns. If most cases have unique covariate patterns, as is often the situation when covariates are not categorical, the goodness-of-fit tests will not have a chi-square distribution, since the expected values for the cells will be small (see Hosmer and Lemeshow, 2000).

Another consideration when evaluating the goodness of fit of a model is determining whether to base the goodness-of-fit tests only on covariate patterns defined by variables in the model or to include additional variables that define the table. For example, if respondents are cross-classified on the basis of *gender*, *age group*, and *degree* but if the final model included only *gender* and *degree*, the values of the goodness-of-fit tests would differ depending on whether the covariate patterns were combinations of only *degree* and *gender* or combinations of *degree*, *gender*, and *age group*. That's because the saturated models are different in the two situations.

You can specify the variables that define the saturated model in the Statistics dialog box. This feature also allows you to calculate the change in –2 log-likelihood when a set of variables is removed from a model, since you can define the saturated model to be the same for both.

Examining the Residuals

As in other statistical procedures, the examination of residuals and other diagnostic statistics plays an important role in the evaluation of the suitability of a particular model. You can save predicted probabilities from the Multinomial Logistic Regression procedure. You can use the Logit or Logistic Regression procedure to calculate and examine other diagnostic measures for pairs of groups. Begg and Gray (1984) show that estimates of the regression coefficients from separate-fit logistic models are close to those from the multinomial fit.

Pseudo-R-square Measures

In linear models, the R^2 statistic represents the proportion of variability in the dependent variable that can be explained by the independent variables. It is easily calculated and interpreted. For logistic regression models, an easily interpretable measure of the strength of the relationship between the dependent variable and the independent variables is not available, although a variety of measures have been proposed.

Figure 3-12
Pseudo-R-square statistics

Cox and Snell	.033
Nagelkerke	.039
McFadden	.017

As shown in Figure 3-12 above, SPSS calculates three such pseudo-R^2 statistics.

Cox and Snell R^2	$R^2_{CS} = 1 - \left(\dfrac{L(\mathbf{B}^{(0)})}{L(\hat{\mathbf{B}})} \right)^{\frac{2}{n}}$
Nagelkerke's R^2	$R^2_{N} = \dfrac{R^2_{CS}}{1 - L(\mathbf{B}^{(0)})^{2/n}}$
McFadden's R^2	$R^2_{M} = 1 - \left(\dfrac{L(\hat{\mathbf{B}})}{L(\mathbf{B}^{(0)})} \right)$

$L(\hat{\mathbf{B}})$ is the log-likelihood function for the model with the estimated parameters, $L(\mathbf{B}^{(0)})$ is the kernel of the log-likelihood of the intercept-only model, and n is the number of cases (sum of all weights). For this example, the values of all of the pseudo-R^2 statistics are small.

Correcting for Overdispersion

Most statistical procedures for categorical data assume what is called multinomial sampling. However, the parameter estimates obtained from most procedures are the same for sampling under other models, such as the Poisson. Occasionally, data show more variability than you would expect, based on the sampling scheme. This is called **overdispersion**. Different causes, such as correlated observations or mixtures of different distributions, can result in overdispersion. It is possible to estimate constants from the data that you can use to correct the variance-covariance matrix of parameter estimates (McCullagh and Nelder, 1989). The Multinomial Logistic Regression procedure can estimate correction factors for overdispersion. The Wald tests are then based on the corrected values.

Automated Variable Selection

As in other regression-like procedures, you can use variable-selection algorithms to help you arrive at a model. As always, the selected model is not "optimal" in any absolute sense, and the significance levels for any tests cannot be interpreted in the usual manner. (For more information, see the chapter "Multiple Linear Regression" in the *SPSS 14.0 Statistical Procedures Companion* [Norušis, 2005].)

Variables can be selected by adding them to a model that contains only the intercept (forward selection) or by removing them from a model that contains all eligible variables (backward elimination). You can also use stepwise versions of these algorithms, in which all variables already in a model are reevaluated for entry or removal at each step. Variables are evaluated for entry based on either the score statistic or the likelihood-ratio test. Variables are evaluated for removal based on either the score statistic or the Wald test.

Hierarchical Variable Entry

The variable selection algorithm in the Multinomial Logistic Regression procedure has the additional feature of allowing constraints on variable entry and removal, based on maintaining the hierarchical structure of a model. A model is hierarchical if, for any interaction in the model, all lower-order interactions and main effects involving those variables are also included. For example, if the interaction *ABC* is in the model, the main effects *A*, *B*, and *C* and the interactions *AB*, *AC*, and *BC* must also be in the model.

There are four different choices for handling the hierarchy issue. Unless you are very familiar with the issues involved and have good theoretical reasons for doing something different, you should use the default. The four options in the Multinomial Logistic Regression Options dialog box are:

- **Treat covariates like factors for the purposes of determining hierarchy.** The default. Makes no distinction between treatment of factors and covariates, and requires that any lower-order terms implied by higher-order terms be included if higher-order terms are included. For example, *X**X cannot be in the model without *X* also being included.

- **Consider only factorial terms for determining hierarchy; any terms with covariates can be entered at any time.** Requires maintenance of hierarchy only for terms involving factors.

- **Within covariate effects, consider only factorial terms for determining hierarchy.** Requires lower-order effects formally "contained" in higher-order effects to be entered before the higher-order effects. This differs from the default option in that for a term involving a covariate to be contained in another term, the two must have identical covariate effects rather than one being implied by the other. For example, *X* is not contained in *X**X, because *X* is not identical to *X**X.

- **Hierarchically constrain entry and removal of terms.** Deselecting this option allows any terms to be added or removed at any time.

For more detailed discussion and examples of the rules, see the *SPSS Command Syntax Reference*, available from the SPSS Help menu. In the NOMREG command, see the discussion of the RULE keyword on the STEPWISE subcommand and the discussion of hierarchy and containment rules following the list of keywords. In order, the four options listed above are denoted in command syntax as RULE(SINGLE), RULE(SFACTOR), RULE(CONTAINMENT), and RULE(NONE).

Specifying the Analysis

Gender and occupational prestige are just two of the many variables that can be related to perception of money spent on space funding. Consider these additional variables: *income3* (income in three categories) and *politics* (1 = strong Democrat, 3 = strong Republican, 2 = neither). To build a model using forward stepwise variable selection, recall the Multinomial Logistic Regression dialog box and select:

▸ Factor(s): male, income3, prestige3, politics

Model...
  Specify Model
  ⊙ Custom/Stepwise

  ▸ Stepwise Terms: male, income3, prestige3, politics

  Stepwise Method: Forward stepwise

Statistics...
  Model
  ☑ Step summary
  ☑ Model fitting information
  ☑ Information criteria
  Parameters
  ☑ Likelihood ratio tests

Step Output

In forward variable selection, the first model contains only the intercept. At the first step, the variable that causes the largest change in –2 log-likelihood is entered, provided that it meets the entry criteria. Figure 3-13 shows that the addition of *income3* results in the largest change in –2 log-likelihood (54.768) and that the observed significance level for the change is less than 0.05, the default entry criterion. The next variable to enter is *male*, then *politics*, and finally *prestige*. After each variable is entered, the remaining variables in the model are checked to make sure that the observed significance level for the removal statistic is less than the specified value (the default is 0.10). Variable entry stops when no more variables meet entry or removal criteria.

Figure 3-13
Variable-entry statistics

Model		Action	Effect(s)	Model Fitting Criteria			Effect Selection Tests		
				AIC	BIC	-2 Log Likelihood	Chi-Square[1,2]	df	Sig.
Step 0	0	Entered	Intercept	544.317	555.720	540.317	.		
Step 1	1	Entered	income3	497.549	531.756	485.549	54.768	4	.000
Step 2	2	Entered	male	462.711	508.321	446.711	38.838	2	.000
Step 3	3	Entered	politics	448.698	517.112	424.698	22.014	4	.000
Step 4	4	Entered	prestige3	440.597	531.816	408.597	16.101	4	.003

Stepwise Method: Forward Stepwise

[1] The chi-square for entry is based on the likelihood ratio test.

[2] The chi-square for removal is based on the likelihood ratio test.

Information Criteria

As you add variables to a model, –2 log-likelihood always decreases. This doesn't mean that models with more variables are better than models with fewer variables, since, as variables are added, you capitalize on chance associations that may be present in a particular data set. A model estimated from a sample usually fits the sample better than it will fit the population from which the sample is selected. That's why you want a model that fits the data well, using a small number of variables.

Since –2 log-likelihood values always decrease, several measures have been suggested for comparing models with different numbers of variables. The Akaike information criterion (AIC) adjusts for different numbers of predictor variables by adding two times the number of independent parameters to the –2 log-likelihood:

$$AIC = -2 \log likelihood + (2 \times number of parameters)$$

Schwarz (1978) and Raftery (1986) suggested the use of the Bayesian information criterion (BIC). It is computed as

$$BIC = -2 \log likelihood + \log(n) \times df$$

where *n* is the total sample size and *df* is the degrees of freedom for the model. If you scan the values of the AIC and BIC shown in Figure 3-13, you see that the smallest value of AIC is for the model with all four predictor variables, while the smallest value of BIC is for the model with two predictor variables. It's usually the case that the BIC criterion favors models with a smaller number of variables than does the AIC criterion.

Likelihood-Ratio Tests for Individual Effects

Figure 3-14 contains the likelihood-ratio tests for variables in the final model. These are calculated by removing each effect in turn from a model that contains all of the effects in the final model. If you compare Figure 3-14 to Figure 3-13, you see that the likelihood-ratio chi-square value is the same in both figures for only *prestige3*. That's because *prestige3* is the last variable entered into the model. For the other variables, the values in Figure 3-13 are calculated as changes from the model at the previous step.

Figure 3-14
Likelihood-ratio tests for effects in the model

Effect	Model Fitting Criteria			Likelihood Ratio Tests		
	AIC of Reduced Model	BIC of Reduced Model	-2 Log Likelihood of Reduced Model	Chi-Square	df	Sig.
Intercept	440.597	531.816	408.597[1]	.000	0	.
male	476.680	556.497	448.680	40.083	2	.000
income3	467.485	535.900	443.485	34.888	4	.000
prestige3	448.698	517.112	424.698	16.101	4	.003
politics	454.299	522.713	430.299	21.701	4	.000

The chi-square statistic is the difference in -2 log-likelihoods between the final model and a reduced model. The reduced model is formed by omitting an effect from the final model. The null hypothesis is that all parameters of that effect are 0.

[1] This reduced model is equivalent to the final model because omitting the effect does not increase the degrees of freedom.

Matched Case-Control Studies

Logistic regression models can be used to analyze data from several different experimental designs. For example, you can take a single random sample of people and then determine who experienced the event of interest (cases) and who did not (controls), or you can take two independent samples, one of cases and one of controls. Sampling cases and controls separately is particularly useful for rare events because you can be sure that you will have enough events to analyze. For both situations, the coefficients for the independent variables from the usual logistic regression analysis procedures are correct. However, for the two-sample situation, you will not be able to estimate the probability of the event in the population for various combinations of risk factors unless you know the sampling fractions for the cases and controls and adjust the intercept parameter accordingly.

Another type of experimental design that can be analyzed with logistic regression is the **matched case-control** study. In a matched case-control study, each case is paired with one or more controls that have the same values for preselected risk factors (matched factors), such as age or gender. For each case and control, information is also gathered about other possible risk factors (unmatched variables). The advantage of such a design is that differences between cases and controls with respect to an event are then attributed to the unmatched risk factors.

The SPSS Multinomial Logistic Regression procedure can be used to analyze data from matched case-control studies in which each case is paired with a single control. These are sometimes called **1–1 matched case-control studies**. The SPSS Cox Regression procedure can be used to analyze data when each case has one or more controls (see Chapter 8). If you use the Cox Regression procedure, you don't have to create the difference variables.

The Model

Consider a 1–1 matched design in which there are K matched pairs of cases and controls. The risk of an individual experiencing the event has two components: the risk associated with the matched variables and the risk associated with the unmatched variables. For a particular individual, the logit for experiencing the event can be written as

$$\log(\text{odds of event}) = \alpha_k + \sum_{i=1}^{p} B_i X_i$$

where α_k is the risk for the kth pair, based on the values of the matched variables; X_1 to X_p are the values of the unmatched independent variables; and B_i is the logistic regression coefficient for the ith unmatched independent variable.

The log of the ratio of the odds that a case will experience the event to the odds that the corresponding control will experience the event can be written as

$$\log\left(\frac{\text{odds of event for case}}{\text{odds of event for control}}\right) = \sum_{i=1}^{p} B_i D_i$$

where B_i is the coefficient for the ith nonmatched independent variable and D_i is the difference in values between the case and its matched control. The logistic regression coefficients are interpreted similarly for matched and unmatched logistic regression analyses.

Creating the Difference Variables

To use the SPSS Multinomial Logistic Regression procedure to analyze data from 1–1 matched designs, the data file must be structured in a special way to reflect the pairing. For the paired analysis, the number of cases must be equal to the number of matched pairs, and the ***variables must be the differences in values between the case and the control.*** Categorical variables must first be transformed and then differenced.

As an example, consider the data from Hosmer and Lemeshow (2000). These data are copyrighted by John Wiley and Sons and are available at:

http://www-unix.oit.umass.edu/~statdata/statdata/data/lowbwtm11.dat

You can copy the data to a text file and then import it into SPSS (File menu, Read Text Data). There are 56 pairs of mothers—one from each pair gave birth to a low-birth-weight baby and the other did not. The women are matched on age. The variables that we will consider are *lwt*, which is the last weight prior to pregnancy; *age*; *race* (1 = *white*, 2 = *black*, 3 = *other*); *smoke* (smoking during pregnancy, 0 = *no*, 1 = *yes*); *ptd* (previous preterm delivery, 0 = *no*, 1 = *yes*); and *ui* (uterine irritability, 0 = *no*, 1 = *yes*).

Table 3-1
Data values and differences for a case-control pair

	LOW	LWT	AGE	RACE	SMOKE	PTD	UI	RACE1	RACE2
Case	1	101	14	3	1	1	0	0	1
Control	0	135	14	1	0	0	0	0	0
Difference		–34		X	1	1	0	0	1

Table 3-1 shows the data values and differences for one of the matched pairs. Calculating differences is straightforward except for categorical variables with more than two categories. You can use the Compute Variable dialog box (Transform menu, Compute) to calculate the differences.

Transforming Categorical Variables

The SPSS Multinomial Logistic Regression procedure automatically transforms all factor variables. However, to run a matched case-control analysis, you must create the new variables used to represent a categorical variable and find the differences between these new variables.

Table 3-2
Indicator coding of race

	RACE1	**RACE2**
White	0	0
Black	1	0
Other	0	1

Consider a simple example. *Race* is a categorical variable with three values, so you will need two variables to represent it. If you use indicator coding with 1 (*white*) as the reference category, you must create two new variables, say *race1* and *race2*. *Race1* will be coded 1 for *black*; 0, otherwise. *Race2* will be coded 1 for *other*; 0, otherwise. Table 3-2 shows the representation of the three race categories. For each case, you must compute the values of *race1* and *race2*. Then, for each case-control pair, you must calculate the difference between *race1* and *race2*, as shown in Table 3-1. The same procedure must be followed for interaction terms. Interaction variables must be created first and then differenced.

The Data File

To analyze a matched case-control study in SPSS Multinomial Logistic Regression, you must have a data file in which each observation consists of the differences for the unmatched variables between a case and the corresponding control. You may also want to keep the values for the matched variables on the record so that you can use them in interaction terms. The dependent variable must be set to a constant for all observations.

Specifying the Analysis

There are tricks to specifying the analysis for a matched case-control design:

■ The dependent variable must have the same value for all cases.

■ All of the difference variables must be identified as *covariates*.

■ In the Multinomial Logistic Regression Model dialog box, the intercept must be suppressed.

■ Matched variables cannot be entered into the model as main effects, since their difference values are 0 for all cases. However, matched variables in their undifferenced form can be included in interaction terms.

To run the analysis, from the menus choose:

Analyze
 Regression
  Multinomial Logistic...

  ▶ Dependent: constant

  ▶ Covariate(s): smoke, ui, ptd, race1, race2, lwt

Model...
 Include intercept in model (deselect)

Statistics...
 Parameters
  ☑ Estimates

Examining the Results

Figure 3-15
Parameter estimates

constant		B	Std. Error	Wald	df	Sig.	Exp(B)	95% Confidence Interval for Exp(B)	
								Lower Bound	Upper Bound
1.00	smoke	1.348	.568	5.631	1	.018	3.849	1.264	11.715
	ui	1.032	.664	2.418	1	.120	2.808	.764	10.316
	ptd	1.563	.706	4.897	1	.027	4.774	1.196	19.061
	race1	.861	.643	1.793	1	.181	2.365	.671	8.338
	race2	.469	.604	.603	1	.438	1.598	.489	5.219
	lwt	-.009	.009	1.051	1	.305	.991	.973	1.009

The parameter estimates for the case-control data set are shown in Figure 3-15. Although the original names of the variables are used, all of the variables represent differences between the case and control pairs.

Figure 3-16
Likelihood-ratio tests

Effect	Model Fitting Criteria	Likelihood Ratio Tests		
	-2 Log Likelihood of Reduced Model	Chi-Square	df	Sig.
smoke	64.422	6.599	1	.010
ui	60.488	2.665	1	.103
ptd	64.030	6.207	1	.013
race1	59.680	1.857	1	.173
race2	58.440	.617	1	.432
lwt	58.970	1.147	1	.284

The chi-square statistic is the difference in -2 log-likelihoods between the final model and a reduced model. The reduced model is formed by omitting an effect from the final model. The null hypothesis is that all parameters of that effect are 0.

The likelihood-ratio tests for each of the effects are shown in Figure 3-16. The coefficients for all of the variables except *lwt* are positive. Only *smoke* and *ptd* are significantly associated with the low-birth-weight outcome. Smoking increases the odds of low birth weight by a factor of 3.8; prior preterm delivery increases the odds of low birth weight by a factor of 4.8. These factors are found in the column labeled *Exp(B)* in Figure 3-15.

Figure 3-17
Model-fitting summary

Model	Model Fitting Criteria	Likelihood Ratio Tests		
	-2 Log Likelihood	Chi-Square	df	Sig.
Null	77.632			
Final	57.823	19.809	6	.003

Figure 3-17 shows that the independent variables are significantly associated with the outcome. The change in –2 log-likelihood is significant when the model with all of the variables is compared to the model with no independent variables and no intercept. You cannot use the goodness-of-fit tests to evaluate how well the model fits, since you don't have multiple observations at each combination of values of the differenced variables. All pairs have different combinations of values. Hosmer and Lemeshow (2000) present a detailed discussion of fitting models to these data, as well as assessing the fit of the model.

Ordinal Regression

Many variables of interest are ordinal. That is, you can rank the values, but the real distance between categories is unknown. Diseases are graded on scales from *least severe* to *most severe*. Survey respondents choose answers on scales from *strongly agree* to *strongly disagree*. Students are graded on scales from *A* to *F*.

You can use ordinal categorical variables as predictors, or factors, in many statistical procedures, such as linear regression. However, you have to make difficult decisions. Should you forget the ordering of the values and treat your categorical variables as if they are nominal? Should you substitute some sort of scale (for example, numbers 1 to 5) and pretend the variables are interval? Should you use some other transformation of the values hoping to capture some of that extra information in the ordinal scale?

When your dependent variable is ordinal, you also face a quandary. You can forget about the ordering and fit a multinomial logit model that ignores any ordering of the values of the dependent variable. You fit the same model if your groups are defined by color of car driven or severity of a disease. You estimate coefficients that capture differences between all possible pairs of groups. Or you can apply a model that incorporates the ordinal nature of the dependent variable. However, keep in mind that even when the categories of the dependent variable can be ordered, that doesn't mean that an ordinal model is necessarily the most appropriate model, especially if categories are ordered on more than one dimension, such as strength of opinion and direction, or if categories can be ordered in different ways. (See, for example, Miller and Volker (1985)).

The SPSS Ordinal Regression procedure, or PLUM (**Po**lytomous **U**niversal **M**odel), is an extension of the general linear model to ordinal categorical data. You can specify five link functions as well as scaling parameters. The procedure can be used to fit heteroscedastic probit and logit models.

Fitting an Ordinal Logit Model

Before delving into the formulation of ordinal regression models as specialized cases of the general linear model, let's consider a simple example. To fit a binary logistic regression model, you estimate a set of regression coefficients that predict the probability of the outcome of interest. The same logistic model can be written in different ways. The version that shows what function of the probabilities results in a linear combination of parameters is

$$\ln\left(\frac{\text{prob(event)}}{(1 - \text{prob(event)})}\right) = \beta_0 + \beta_1 X_1 + \beta_2 X_2 + \dots + \beta_k X_k$$

The quantity to the left of the equals sign is called a **logit**. It's the log of the odds that an event occurs. (The odds that an event occurs is the ratio of the number of people who experience the event to the number of people who do not. This is what you get when you divide the probability that the event occurs by the probability that the event does not occur, since both probabilities have the same denominator and it cancels, leaving the number of events divided by the number of non-events.) The coefficients in the logistic regression model tell you how much the logit changes based on the values of the predictor variables.

When you have more than two events, you can extend the binary logistic regression model, as described in Chapter 3. For ordinal categorical variables, the drawback of the multinomial regression model is that the ordering of the categories is ignored.

Modeling Cumulative Counts

You can modify the binary logistic regression model to incorporate the ordinal nature of a dependent variable by defining the probabilities differently. Instead of considering the probability of an individual event, you consider the probability of that event and all events that are ordered before it.

Consider the following example: A random sample of Vermont voters was polled. The voters were asked to rate their satisfaction with the criminal justice system in the state (Doble, 1999). They rated judges on the scale: *Poor* (1), *Only fair* (2), *Good* (3), and *Excellent* (4). Each voter also indicated whether he or she or anyone in his or her family was a crime victim in the last three years. You want to model the relationship between their rating and having a crime victim in the household.

Defining the Event

In ordinal logistic regression, the event of interest is observing a particular score *or less*. For the rating of judges, you model the following odds:

θ_1 = prob(score of 1) / prob(score greater than 1)
θ_2 = prob(score of 1 or 2) / prob(score greater than 2)
θ_3 = prob(score of 1, 2, or 3) / prob(score greater than 3)

The last category doesn't have an odds associated with it since the probability of scoring up to and including the last score is 1.

All of the odds are of the form:

θ_j = prob(score $\leq j$) / prob(score $> j$)

You can also write the equation as

θ_j = prob(score $\leq j$) / (1 − prob(score $\leq j$)),

since the probability of a score greater than j is 1 minus probability of a score less than or equal to j.

Ordinal Model

The ordinal logistic model for a single independent variable is then

$\ln(\theta_j) = \alpha_j - \beta X$

where j goes from 1 to the number of categories minus 1.

It is not a typo that there is a minus sign before the coefficients for the predictor variables, instead of the customary plus sign. That is done so that larger coefficients indicate an association with larger scores. When you see a positive coefficient in SPSS output for a dichotomous factor, you know that higher scores are more likely for the first category. A negative coefficient tells you that lower scores are more likely. For a continuous variable, a positive coefficient tells you that as the values of the variable increase, the likelihood of larger scores increases. An association with higher scores means smaller cumulative probabilities for lower scores, since they are less likely to occur.

Each logit has its own α_j term but the same coefficient β. That means that the effect of the independent variable is the same for different logit functions. That's an assumption you have to check. That's also the reason the model is also called the proportional odds model. The α_j terms, called the threshold values, often aren't of much interest. Their values do not depend on the values of the independent variable for a particular case. They are like the intercept in a linear regression, except that each logit has its own intercept. They're used in the calculations of predicted values. From the previous equations, you also see that combining adjacent scores into a single category won't change the results for the groups that aren't involved in the merge. That's a desirable feature.

Examining Observed Cumulative Counts

Before you start building any model, you should examine the data. Figure 4-1 is a cumulative percentage plot of the ratings, with separate curves for those whose households experienced crime and those who didn't. The lines for those who experienced crime are above the lines for those who didn't. Figure 4-1 also helps you visualize the ordinal regression model. It models a function of those two curves.

Consider the rating *Poor*. A larger percentage of crime victims than non-victims chose this response. (Because it is the first response, the cumulative percentage is just the observed percentage for the response.) As additional percentages are added (the cumulative percentage for *Only fair* is the sum of *Poor* and *Only fair),* the cumulative percentages for the crime victim households remain larger than for those without crime. It's only at the end, when both groups must reach 100%, that they must join. Because the victims assign lower scores, you expect to see a negative coefficient for the predictor variable, *hhcrime* (household crime experience).

Figure 4-1

Plot of observed cumulative percentages

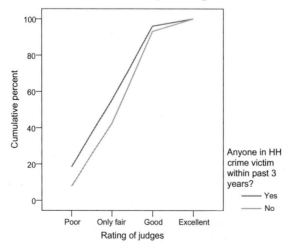

Specifying the Analysis

To fit the cumulative logit model, open the file *vermontcrime.sav* and from the menus choose:

Analyze
  Regression
    Ordinal...

▶ Dependent: rating
▶ Factors: hhcrime

Options...
Link: Logit

Output...
  Display
    ☑ Goodness of fit statistics
    ☑ Summary statistics
    ☑ Parameter estimates
    ☑ Cell information
    ☑ Test of Parallel Lines
  Saved Variables
    ☑ Estimated response probabilities

Parameter Estimates

Figure 4-2 contains the estimated coefficients for the model. The estimates labeled *Threshold* are the α_j s, the intercept equivalent terms. The estimates labeled *Location* are the ones you're interested in. They are the coefficients for the predictor variables. The coefficient for *hhcrime* (coded 1 = *yes*, 2 = *no*), the independent variable in the model, is –0.633. As is always the case with categorical predictors in models with intercepts, the number of coefficients displayed is one less than the number of categories of the variable. In this case, the coefficient is for the value of 1. Category 2 is the reference category and has a coefficient of 0.

The coefficient for those whose household experienced crime in the past three years is negative, as you expected from Figure 4-1. That means it's associated with poorer scores on the rankings of judges. If you calculate $e^{-\beta}$, that's the ratio of the odds for lower to higher scores for those experiencing crime and those not experiencing crime. In this example, $\exp(0.633) = 1.88$. This ratio stays the same over all of the ratings.

The Wald statistic is the square of the ratio of the coefficient to its standard error. Based on the small observed significance level, you can reject the null hypothesis that it is zero. There appears to be a relationship between household crime and ratings of judges. For any rating level, people who experience crime score judges lower than those who don't experience crime.

Figure 4-2
Parameter estimates

		Estimate	Std. Error	Wald	df	Sig.	95% Confidence Interval	
							Lower Bound	Upper Bound
Threshold	[rating = 1]	-2.392	.152	248.443	1	.000	-2.690	-2.095
	[rating = 2]	-.317	.091	12.146	1	.000	-.495	-.139
	[rating = 3]	2.593	.172	228.287	1	.000	2.257	2.930
Location	[hhcrime=1]	-.633	.232	7.445	1	.006	-1.088	-.178
	[hhcrime=2]	0[1]	.	.	0	.	.	.

Link function: Logit.

[1]. This parameter is set to zero because it is redundant.

Testing Parallel Lines

When you fit an ordinal regression, you assume that the relationships between the independent variables and the logits are the same for all the logits. That means that the results are a set of parallel lines or planes—one for each category of the outcome variable. You can check this assumption by allowing the coefficients to vary, estimating them, and then testing whether they are all equal.

The result of the test of parallelism is in Figure 4-3. The row labeled *Null Hypothesis* contains –2 log-likelihood for the constrained model, the model that assumes the lines are parallel. The row labeled *General* is for the model with separate lines or planes. You want to know whether the general model results in a sizeable improvement in fit from the null hypothesis model.

The entry labeled *Chi-Square* is the difference between the two –2 log-likelihood values. If the lines or planes are parallel, the observed significance level for the change should be large, since the general model doesn't improve the fit very much. The parallel model is adequate. You don't want to reject the null hypothesis that the lines are parallel. From Figure 4-3, you see that the assumption is plausible for this problem. If you do reject the null hypothesis, it is possible that the link function selected is incorrect for the data or that the relationships between the independent variables and logits are not the same for all logits.

Figure 4-3

Test of parallel lines

Test of Parallel Lines[1]

Model	-2 Log Likelihood	Chi-Square	df	Sig.
Null Hypothesis	30.793			
General	28.906	1.887	2	.389

The null hypothesis states that the location parameters (slope coefficients) are the same across response categories.

[1] Link function: Logit.

Does the Model Fit?

A standard statistical maneuver for testing whether a model fits is to compare observed and expected values. That is what's done here as well.

Calculating Expected Values

You can use the coefficients in Figure 4-2 to calculate cumulative predicted probabilities from the logistic model for each case:

$$\text{prob(event } j) = 1 / (1 + e^{-(\alpha_j - \beta x)})$$

Remember that the events in an ordinal logistic model are not individual scores but cumulative scores. First, calculate the predicted probabilities for those who didn't experience household crime. That means that β is 0, and all you have to worry about are the intercept terms.

$$\text{prob(score 1)} = 1 / (1 + e^{2.392}) = 0.0838$$
$$\text{prob(score 1 or 2)} = 1 / (1 + e^{0.317}) = 0.4214$$
$$\text{prob(score 1 or 2 or 3)} = 1 / (1 + e^{-2.59}) = 0.9302$$
$$\text{prob(score 1 or 2 or 3 or 4)} = 1$$

From the estimated cumulative probabilities, you can easily calculate the estimated probabilities of the individual scores for those whose households did not experience crime. You calculate the probabilities for the individual scores by subtraction, using the formula:

$$\text{prob(score} = j) = \text{prob(score less than or equal to } j) - \text{prob(score less than } j).$$

The probability for score 1 doesn't require any modifications. For the remaining scores, you calculate the differences between cumulative probabilities:

prob(score = 2) = prob(score = 1 or 2) – prob(score = 1) = 0.3376
prob(score = 3) = prob(score 1, 2, 3) – prob(score 1, 2) = 0.5088
prob(score = 4) = 1 – prob(score 1, 2, 3) = 0.0698

You calculate the probabilities for those whose households experienced crime in the same way. The only difference is that you have to include the value of β in the equation. That is,

prob(score = 1) = $1 / (1 + e^{(2.392 - 0.633)})$ = 0.1469
prob(score = 1 or 2) = $1 / (1 + e^{(0.317 - 0.633)})$ = 0.5783
prob(score = 1, 2, or 3) = $1 / (1 + e^{(-2.593 - 0.633)})$ = 0.9618
prob(score = 1, 2, 3, or 4) = 1

Of course, you don't have to do any of the actual calculations, since SPSS will do them for you. In the Options dialog box, you can ask that the predicted probabilities for each score be saved.

Figure 4-4 gives the predicted probabilities for each cell. The output is from the Means procedure with the saved predicted probabilities (*EST1_1*, *EST2_1*, *EST3_1*, and *EST4_1*) as the dependent variables and *hhcrime* as the factor variable. All cases with the same value of *hhcrime* have the same predicted probabilities for all of the response categories. That's why the standard deviation in each cell is 0. For each rating, the estimated probabilities for everybody combined are the same as the observed marginals for the rating variable.

Figure 4-4
Estimated response probabilities

Anyone in HH crime victim within past 3 years?		Estimated Cell Probability for Response Category: 1	Estimated Cell Probability for Response Category: 2	Estimated Cell Probability for Response Category: 3	Estimated Cell Probability for Response Category: 4
Yes	Mean	.1469	.4316	.3833	.0382
	N	76	76	76	76
	Std. Deviation	.00000	.00000	.00000	.00000
No	Mean	.0838	.3378	.5089	.0696
	N	490	490	490	490
	Std. Deviation	.00000	.00000	.00000	.00000
Total	Mean	.0922	.3504	.4920	.0654
	N	566	566	566	566
	Std. Deviation	.02154	.03202	.04284	.01071

For each rating, the estimated odds of the cumulative ratings for those who experience crime divided by the estimated odds of the cumulative ratings for those who didn't experience crime is $e^{-\beta} = 1.88$. For the first response, the odds ratio is

$$\frac{0.1469/(1 - 0.1469)}{0.0838/(1 - 0.0838)} = 1.88$$

For the cumulative probability of the second response, the odds ratio is

$$\frac{(0.1469 + 0.4316)/(1 - 0.1469 - 0.4316)}{(0.0838 + 0.3378)/(1 - 0.0838 - 0.3378)} = 1.88$$

Comparing Observed and Expected Counts

You can use the previously estimated probabilities to calculate the number of cases you expect in each of the cells of a two-way crosstabulation of rating and crime in the household. You multiply the expected probabilities for those without a history by 490, the number of respondents who didn't report a history. The expected probabilities for those with a history are multiplied by 76, the number of people reporting a household history of crime. These are the numbers you see in Figure 4-5 in the row labeled *Expected*. The row labeled *Observed* is the actual count.

The Pearson residual is a standardized difference between the observed and predicted values:

$$\text{Pearson residual} = \frac{O_{ij} - E_{ij}}{\sqrt{n_i \hat{p}_{ij}(1 - \hat{p}_{ij})}}$$

Figure 4-5
Cell information

Frequency

Anyone in HH crime victim within past 3 years?		Rating of judges			
		Poor	Only fair	Good	Excellent
Yes	Observed	14	28	31	3
	Expected	11.16	32.800	29.134	2.903
	Pearson Residual	.919	-1.112	.440	.058
No	Observed	38	170	248	34
	Expected	41.05	165.501	249.4	34.094
	Pearson Residual	-.497	.430	-.123	-.017

Link function: Logit.

Goodness-of-Fit Measures

From the observed and expected frequencies, you can compute the usual Pearson and Deviance goodness-of-fit measures. The Pearson goodness-of-fit statistic is

$$\chi^2 = \Sigma\Sigma\frac{(O_{ij} - E_{ij})^2}{E_{ij}}$$

The deviance measure is

$$D = 2\Sigma\Sigma O_{ij}\ln\left(\frac{O_{ij}}{E_{ij}}\right)$$

Both of the goodness-of-fit statistics should be used only for models that have reasonably large expected values in each cell. If you have a continuous independent variable or many categorical predictors or some predictors with many values, you may have many cells with small expected values. SPSS warns you about the number of empty cells in your design. In this situation, neither statistic provides a dependable goodness-of-fit test.

If your model fits well, the observed and expected cell counts are similar, the value of each statistic is small, and the observed significance level is large. You reject the null hypothesis that the model fits if the observed significance level for the goodness-of-fit statistics is small. Good models have large observed significance levels. In Figure 4-6, you see that the goodness-of-fit measures have large observed significance levels, so it appears that the model fits.

Figure 4-6
Goodness-of-fit statistics

	Chi-Square	df	Sig.
Pearson	1.902	2	.386
Deviance	1.887	2	.389

Link function: Logit.

Including Additional Predictor Variables

A single predictor variable example makes explaining the basics easier, but real problems almost always involve more than one predictor. Consider what happens when additional factor variables—such as *sex*, *age2* (two categories), and *educ5* (five categories)—are included as well.

Recall the Ordinal Regression dialog box and select:

▶ Dependent: rating
▶ Factors: hhcrime, sex, age2, educ5

Options...
　Link: Logit

Output...
　Display
　☑ Goodness of fit statistics
　☑ Summary statistics
　☑ Parameter estimates
　☑ Test of Parallel Lines
　Saved Variables
　☑ Predicted category

The dimensions of the problem have quickly escalated. You've gone from 8 cells, defined by the four ranks and two crime categories, to 160 cells. The number of cases with valid values for all of the variables is 536, so cells with small observed and predicted frequencies will be a problem for the tests that evaluate the goodness of fit of the model. That's why the warning in Figure 4-7 appears.

Figure 4-7
Warning for empty cells

There are 44 (29.7%) cells (i.e., dependent variable levels by combinations of predictor variable values) with zero frequencies.

Overall Model Test

Before proceeding to examine the individual coefficients, you want to look at an overall test of the null hypothesis that the location coefficients for all of the variables in the model are 0. You can base this on the change in –2 log-likelihood when the variables are added to a model that contains only the intercept. The change in likelihood function has a chi-square distribution even when there are cells with small observed and predicted counts.

From Figure 4-8, you see that the difference between the two log-likelihoods—the chi-square—has an observed significance level of less than 0.0005. This means that you can reject the null hypothesis that the model without predictors is as good as the model with the predictors.

Figure 4-8
Model-fitting information

Model	-2 Log Likelihood	Chi-Square	df	Sig.
Intercept Only	322.784			
Final	288.600	34.183	7	.000

Link function: Logit.

You also want to test the assumption that the regression coefficients are the same for all four categories. If you reject the assumption of parallelism, you should consider using multinomial regression, which estimates separate coefficients for each category. Since the observed significance level in Figure 4-9 is large, you don't have sufficient evidence to reject the parallelism hypothesis.

Figure 4-9
Test of parallelism

Model	-2 Log Likelihood	Chi-Square	df	Sig.
Null Hypothesis	288.600			
General	276.115	12.485	14	.567

The null hypothesis states that the location parameters (slope coefficients) are the same across response categories.

Examining the Coefficients

From the observed significance levels in Figure 4-10, you see that sex, education, and household history of crime are all related to the ratings. They all have negative coefficients. Men (code 1) are less likely to assign higher ratings than women, people with less education are less likely to assign higher ratings than people with graduate education (code 5), and persons whose households have been victims of crime are less likely to assign higher ratings than those in crime-free households. Age doesn't appear to be related to the rating.

Figure 4-10

Parameter estimates for the model

		Estimate	Std. Error	Wald	df	Sig.
Threshold	[rating = 1]	-3.630	.335	117.579	1	.000
	[rating = 2]	-1.486	.302	24.265	1	.000
	[rating = 3]	1.533	.311	24.378	1	.000
Location	[hhcrime=1]	-.643	.238	7.318	1	.007
	[hhcrime=2]	0[1]	.	.	0	.
	[sex=1]	-.424	.163	6.758	1	.009
	[sex=2]	0[1]	.	.	0	.
	[age2=0]	.076	.176	.186	1	.666
	[age2=1]	0[1]	.	.	0	.
	[educ5=1]	-1.518	.389	15.198	1	.000
	[educ5=2]	-1.256	.288	19.004	1	.000
	[educ5=3]	-.941	.310	9.188	1	.002
	[educ5=4]	-.907	.302	9.015	1	.003
	[educ5=5]	0[1]	.	.	0	.

Link function: Logit.

[1]. This parameter is set to zero because it is redundant.

Measuring Strength of Association

There are several R^2-like statistics that can be used to measure the strength of the association between the dependent variable and the predictor variables. They are not as useful as the R^2 statistic in regression, since their interpretation is not straightforward. Three commonly used statistics are:

- Cox and Snell R^2

$$R^2{}_{CS} = 1 - \left(\frac{L(\mathbf{B}^{(0)})}{L(\hat{\mathbf{B}})}\right)^{\frac{2}{n}}$$

- Nagelkerke's R^2

$$R^2{}_N = \frac{R^2_{CS}}{1 - L(\mathbf{B}^{(0)})^{2/n}}$$

■ McFadden's R^2

$$R^2{}_M = 1 - \left(\frac{L(\hat{\mathbf{B}})}{L(\mathbf{B}^{(0)})} \right)$$

where $L(\hat{\mathbf{B}})$ is the log-likelihood function for the model with the estimated parameters and $L(\mathbf{B}^{(0)})$ is the log-likelihood with just the thresholds, and n is the number of cases (sum of all weights). For this example, the values of all of the pseudo R-square statistics are small.

Figure 4-11
Pseudo R-square

Cox and Snell	.059
Nagelkerke	.066
McFadden	.027

Link function: Logit.

Classifying Cases

You can use the predicted probability of each response category to assign cases to categories. A case is assigned to the response category for which it has the largest predicted probability. Figure 4-12 is the classification table, which is obtained by crosstabulating *rating* by *pre_1*. (This is sometimes called the confusion matrix.)

Figure 4-12
Classification table

Count

Rating of judges	Predicted Response Category		Total
	Only fair	Good	
Poor	15	36	51
Only fair	42	156	198
Good	33	246	279
Excellent	2	35	37
Total	92	473	565

Of the 198 people who selected the response *Only fair,* only 42 are correctly assigned to the category using the predicted probability. Of the 279 who selected *Good,* 246 are correctly assigned. None of the respondents who selected *Poor* or *Excellent* is correctly assigned. If the goal of your analysis is to study the association between the grouping variable and the predictor variables, the poor classification should not concern you. If your goal is to target marketing or collections efforts, the correct classification rate may be more important.

Generalized Linear Models

The ordinal logistic model is one of many models subsumed under the rubric of generalized linear models for ordinal data. The model is based on the assumption that there is a latent continuous outcome variable and that the observed ordinal outcome arises from discretizing the underlying continuum into *j*-ordered groups. The thresholds estimate these cutoff values.

The basic form of the generalized linear model is

$$\text{link}(\gamma_j) = \frac{\theta_j - [\beta_1 x_1 + \beta_2 x_2 + \ldots + \beta_k x_k]}{\exp(\tau_1 z_1 + \tau_2 z_2 + \ldots + \tau_m z_m)}$$

where γ_j is the cumulative probability for the *j*th category, θ_j is the threshold for the *j*th category, $\beta_1 \ldots \beta_k$ are the regression coefficients, $x_1 \ldots x_k$ are the predictor variables, and k is the number of predictors.

The numerator on the right side determines the **location** of the model. The denominator of the equation specifies the scale. The $\tau_1 \ldots \tau_m$ are coefficients for the scale component, and $z_1 \ldots z_m$ are *m* predictor variables for the scale component (chosen from the same set of variables as the *x*s).

The **scale component** accounts for differences in variability for different values of the predictor variables. For example, if certain groups have more variability than others in their ratings, using a scale component to account for this may improve your model.

Link Function

The link function is the function of the probabilities that results in a linear model in the parameters. It defines what goes on the left side of the equation. It's the link between the random component on the left side of the equation and the systematic component on the right. In the criminal rating example, the link function is the logit function, since the log of the odds results is equal to the linear combination of the parameters. That is,

$$\ln\left(\frac{\text{prob(event)}}{(1 - \text{prob(event)})}\right) = \beta_0 + \beta_1 x_1 + \beta_2 x_2 + \ldots + \beta_k x_k$$

Five different link functions are available in the Ordinal Regression procedure in SPSS. They are summarized in the following table. The symbol γ represents the probability that the event occurs. Remember that in ordinal regression, the probability of an event is redefined in terms of cumulative probabilities.

Function	Form	Typical application
Logit	$\ln\left(\frac{\gamma}{1 - \gamma}\right)$	Evenly distributed categories
Complementary log-log	$\ln(-\ln(1 - \gamma))$	Higher categories more probable
Negative log-log	$-\ln(-\ln(\gamma))$	Lower categories more probable
Probit	$\Phi^{-1}(\gamma)$	Analyses with explicit normally distributed latent variable
Cauchit (inverse Cauchy)	$\tan(\pi(\gamma - 0.5))$	Outcome with many extreme values

If you select the probit link function, you fit the model described in Chapter 5. The observed probabilities are replaced with the value of the standard normal curve below which the observed proportion of the area is found.

Probit and logit models are reasonable choices when the changes in the cumulative probabilities are gradual. If there are abrupt changes, other link functions should be used. The complementary log-log link may be a good model when the cumulative probabilities increase from 0 fairly slowly and then rapidly approach 1. If the opposite is true, namely that the cumulative probability for lower scores is high and the approach to 1 is slow, the negative log-log link may describe the data. If the complementary log-log model describes the probability of an event occurring, the log-log model describes the probability of the event not occurring.

Fitting a Heteroscedastic Probit Model

Probit models are useful for analyzing signal detection data. Signal detection describes the process of detecting an event in the face of uncertainty or "noise." You must decide whether a signal is present or absent. For example, a radiologist has to decide whether a tumor is present or not based on inspecting images. You can model the uncertainty in the decision-making process by asking subjects to report how confident they are in their decision.

You postulate the existence of two normal distributions: one for the probability of detecting a signal when only noise is present and one for detecting the signal when both the signal and the noise are present. The difference between the means of the two distributions is called *d*, a measure of the sensitivity of the person to the signal.

The general probit model is

$$p(Y \leq k|X) = \Phi\left(\frac{c_k - d_nX}{\sigma_s^x}\right)$$

where Y is the dependent variable, such as a confidence rating, with values from 1 to K, X is a 0–1 variable that indicates whether the signal was present or absent, c_k are ordered distances from the noise distribution, d_n is the scaled distance parameter, and σ_s is the standard deviation of the signal distribution. The model can be rewritten as

$$\Phi^{-1}[p(Y \leq k|X)] = \frac{c_k - d_nX}{e^{ax}}$$

where Φ^{-1} is the inverse of the cumulative normal distribution and *a* is the natural log of σ_s. The numerator models the location; the denominator models the scale.

If the noise and signal distributions have different variances, you must include this information in the model. Otherwise, the parameter estimates are biased and inconsistent. Even large sample sizes won't set things right.

Modeling Signal Detection

Consider data reported from a light detection study by Swets, et al. (1961) and discussed by DeCarlo (2003). Data are for a single individual who rated his confidence that a signal was present in 591 trials when the signal was absent and 597 trials when the signal was present.

In Figure 4-13, you see the cumulative distribution of the ratings under the two conditions (signal absent and signal present). The noise curve is above the signal curve, indicating that the low confidence ratings were more frequent when a signal was not present.

Figure 4-13
Plot of cumulative confidence ratings

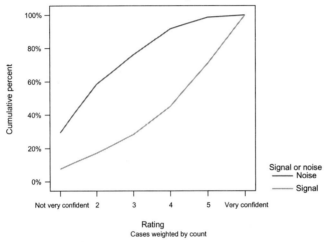

Fitting a Location-Only Model

If you assume that the variances of the noise and signal distributions are equal, you can fit the usual probit model. Open the file *swets.sav*. The data are aggregated. For each possible combination of *signal* and *response*, there is a *count* of the number of times that each response was chosen.

You must weight the data file before proceeding. From the menus choose:

Data
  Weight Cases...

⊙ Weight cases by
  ▶ count

Analyze
 Regression
  Ordinal...

▶ Dependent: response
▶ Covariate(s): signal

Options...
 Link: Probit

Output...
 Display
  ☑ Parameter Estimates
  ☑ Goodness of fit statistics

Examining the Goodness of Fit

Since the model has only 12 cells, and no cells have zero frequencies, you can examine the goodness-of-fit statistics without concern that the expected counts are too small for the chi-square approximation to be valid. From Figure 4-14, you see that the model does not fit well. The observed significance level is less than 0.0005.

Figure 4-14
Goodness-of-fit statistics

	Chi-Square	df	Sig.
Pearson	33.728	4	.000
Deviance	32.972	4	.000

Link function: Probit.

One of the reasons the model may fit poorly is because the variance of the two distributions of responses may be different. You need a model that allows the variance of the underlying variable to vary as a function of one or more of the independent variables.

You can select a model for the standard deviation such that

$$\sigma_i = e^{Z_i \gamma}$$

where Z_i is a vector of covariates selected from the predictor variables.

Fitting a Scale Parameter

To fit a model that allows for different variances in the two groups, you must specify a model for the scale parameters. To fit a model with different variances in the two groups, recall the dialog box and select:

Scale...
   Scale model: signal

Because you have only one independent variable, *signal*, separate variances are estimated for each of the two categories of *signal*. If you have several predictor variables, you can specify a separate model for the scale component.

   The goodness-of-fit statistics in Figure 4-15 indicate that the model fits much better than the location-only model. The variability of the distributions are an important consideration in this problem.

Figure 4-15
Goodness of fit with scale model

	Chi-Square	df	Sig.
Pearson	1.497	3	.683
Deviance	1.482	3	.687

Link function: Probit.

Parameter Estimates

When you fit a model with scale parameters as well as location parameters, parameter estimates for both are displayed.

Figure 4-16
Parameter estimates for model with location and scale parameters

Parameter Estimates

		Estimate	Std. Error	Wald	df	Sig.	95% Confidence Interval	
							Lower Bound	Upper Bound
Threshold	[response = 1]	-.533	.054	98.809	1	.000	-.638	-.428
	[response = 2]	.204	.050	16.979	1	.000	.107	.301
	[response = 3]	.710	.053	182.311	1	.000	.607	.813
	[response = 4]	1.366	.067	414.418	1	.000	1.235	1.498
	[response = 5]	2.294	.113	409.475	1	.000	2.072	2.516
Location	signal	1.519	.096	250.110	1	.000	1.331	1.707
Scale	signal	.348	.063	30.711	1	.000	.225	.472

Link function: Probit.

The threshold values are distances of the response criteria from the mean of the noise distribution. The location parameter estimate is the estimate of the detection parameter, d_n. To convert the scale parameter to an estimate of the ratio of the noise to signal standard deviations, you must compute $e^{-0.348}$, which is 0.71.

Model-Fitting Information

The overall test of the model is shown in Figure 4-17. When there is a scale parameter, the null hypothesis is that *both* the location parameters and the scale parameters are 0. A scale parameter of 0 means that the variances are equal. Based on the small observed significance level, you can reject this composite null hypothesis. Consult DeCarlo (2003) for further discussion of this example and for other examples of using the Ordinal Regression procedure in signal detection.

Figure 4-17
Model-fitting information

Model	-2 Log Likelihood	Chi-Square	df	Sig.
Intercept Only	461.699			
Final	59.261	402.437	2	.000

Link function: Probit.

Probit Regression

How much insecticide does it take to kill a pest? How low does a sale price have to be to induce a consumer to buy a product? In both of these situations, you are concerned with evaluating the potency of a stimulus. In the first example, the stimulus is the amount of insecticide; in the second, it is the sale price of an object. The response is all or none. An insect is either dead or alive; a sale is made or not made. Since all insects and shoppers do not respond in the same way—that is, they have different tolerances for insecticides and sale prices—the statistical problem must be formulated in terms of the proportion responding at each level of the stimulus.

Your data consist of values of one or more stimuli of interest, each having several levels. You expose different groups of individuals to the selected combinations of stimuli. For each combination, you record the number of individuals exposed and the number who respond.

Different mathematical models can be used to express the relationship between the proportion responding and the "dose" of one or more stimuli. In this chapter, you will consider two commonly used models: the probit response model and the logit response model. The SPSS Probit Analysis procedure should be used only for grouped-response dose data. It does not save any diagnostics or predicted values. You should use the SPSS Ordinal Regression procedure, specifying only two categories, for more extensive analyses or if data are not aggregated. An example of a heteroscedastic probit model is discussed in Chapter 4. If your data are aggregated in a form suitable for the Probit Analysis procedure, you will have to reformat the data for the Ordinal Regression procedure, including separate records for counts of the cases that died and the cases that did not.

Probit and Logit Response Models

In probit and logit models, instead of regressing the actual proportion responding to the values of the stimuli, you transform the proportion responding by using either a logit or probit transformation. In a probit transformation, each of the observed proportions is replaced with the value of the standard normal curve below which the observed proportion of the area is found.

For example, if half of the subjects respond at a particular dose, the corresponding probit value is 0, since half of the area in a standard normal curve falls below a z score of 0. If the observed proportion is 0.95, the corresponding probit value is 1.64.

If the logit transformation is used, the observed proportion P is replaced by

$$\ln\left(\frac{P}{1-P}\right)$$

This quantity is called a **logit**. If the observed proportion is 0.5, the logit-transformed value is 0, the same as the probit-transformed value. If the observed proportion is 0.95, the logit-transformed value is 2.94. (In many situations, analyses based on logits and probits give similar results.)

The regression model for the transformed response can be written as

Transformed $P_i = A + BX_i$

where P_i is the observed proportion responding at dose X_i. (In toxicological experiments, the log of the dose is often used instead of the actual dose, since the tolerance distribution for the log dosage is approximately normal.) If there is more than one stimulus variable, terms are added to the model for each of the stimuli. The SPSS Probit Analysis procedure provides maximum-likelihood estimates of the regression coefficients.

Evaluating Insecticides

Finney (1971) presents data showing the effect of a series of doses of rotenone (an insecticide) when sprayed on *Macrosiphoniella sanborni*. Table 5-1 contains the concentration, the number of insects tested at each dose, the proportion dying, and the probit transformation of each of the observed proportions.

Table 5-1
Effects of rotenone

Dose	Number observed	Number dead	Proportion dead	Probit
10.2	50	44	0.88	1.18
7.7	49	42	0.86	1.08
5.1	46	24	0.52	0.05
3.8	48	16	0.33	−0.44
2.6	50	6	0.12	−1.18

Specifying the Analysis

To analyze the data with a probit model, open the file *rotenone.sav* and from the menus choose:

Analyze
  Regression
    Probit...

  ▶ Response Frequency: died
  ▶ Total Observed: total
  ▶ Covariate(s) dose

  Transform: Log base 10
  Model
    ⊙ Probit

Figure 5-1 is a plot of the observed probits against the logs of the concentrations. The relationship between the two variables appears approximately linear. If it isn't linear, the concentrations have to be transformed in some other way to achieve linearity. Otherwise, fitting a linear model isn't a reasonable strategy for analyzing the data.

Figure 5-1
Plot of observed probits against logs of concentration

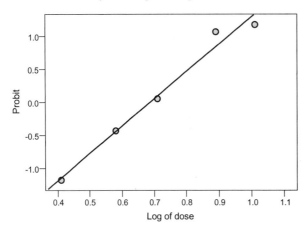

The parameter estimates and standard errors for this example are shown in Figure 5-2. (The algorithm used for estimating the parameters is iterative, so you should check the convergence information table to make sure that the optimal solution was found.)

Figure 5-2
Parameter estimates and standard errors

			Estimate	Std. Error	Z	Sig.	95% Confidence Interval	
							Lower Bound	Upper Bound
PROBIT[1]	Parameter	dose	4.169	.473	8.813	.000	3.242	5.096
		Intercept	-2.859	.347	-8.236	.000	-3.207	-2.512

1. PROBIT model: PROBIT(p) = Intercept + BX (Covariates X are transformed using the base 10 logarithm.)

The regression equation is

$$\text{Probit}(P_i) = -2.86 + 4.17(\log_{10}(\text{dose}_i))$$

To see how well this model fits, look at Figure 5-3, which contains observed and expected frequencies, residuals, and the predicted probability of a response for each of the log dosages.

Figure 5-3
Observed and expected frequencies for each dose

			dose	Number of Subjects	Observed Responses	Expected Responses	Residual	Probability
PROBIT	Number	1	1.010	50	44	45.586	-1.586	.912
		2	.890	49	42	39.330	2.670	.803
		3	.710	46	24	24.845	-.845	.540
		4	.580	48	16	15.816	.184	.329
		5	.410	50	6	6.253	-.253	.125

Based on the Pearson goodness-of-fit test for the model, shown in Figure 5-4, the model appears to fit reasonably well. The Pearson chi-square goodness-of-fit test is calculated as

$$\chi^2 = \sum \frac{(\text{residual}_i)^2}{n_i \hat{P}_i (1 - \hat{P}_i)}$$

where n_i is the number of subjects exposed to dose i, and $\hat{P}_i$ is the predicted proportion responding at dose i. The degrees of freedom are equal to the number of doses minus the number of estimated parameters. In this example, you have five doses and two estimated parameters, so there are three degrees of freedom for the chi-square statistic. Since the observed significance level for the chi-square statistic is large—0.655—there is insufficient reason to doubt the model.

Figure 5-4
Pearson chi-square goodness-of-fit test

		Chi-Square	df[1]	Sig.
PROBIT	Pearson Goodness-of-Fit Test	1.621	3	.655[2]

1. Statistics based on individual cases differ from statistics based on aggregated cases.

2. Since the significance level is greater than .150, no heterogeneity facto is used in the calculation of confidence limits.

When the significance level of the chi-square statistic is small, several explanations are possible. It may be that the relationship between the concentration and the probit is not linear. Or it may be that the relationship is linear, but the spread of the observed points around the regression line is unequal. That is, the data are heterogeneous. If this is the case, a correction must be applied to the estimated variances for each concentration group (see "Confidence Intervals for Effective Dosages" on p. 96).

Confidence Intervals for Effective Dosages

Often, you want to know what the concentration of an agent must be in order to achieve a certain proportion of response. For example, you may want to know what the concentration would have to be in order to kill half of the insects. This is known as the **median lethal dose**. It can be obtained from the previous regression equation by solving for the concentration that corresponds to a probit value of 0. For this example,

$$\log_{10}(\text{median lethal dose}) = 2.86/4.17 = 0.686$$

$$\text{median lethal dose} = 4.85$$

Confidence intervals can be constructed for the median lethal dose as well as for the dose required to achieve any response. The SPSS Probit Analysis procedure calculates 95% intervals for the original concentrations and transformed concentrations required to achieve various levels of response. The values for this example are shown in Figure 5-5.

Figure 5-5
Confidence limits for effective doses

			95% Confidence Limits for dose			95% Confidence Limits for log(dose)[1]		
			Estimate	Lower Bound	Upper Bound	Estimate	Lower Bound	Upper Bound
PROBIT	Prob	.010	1.342	.902	1.740	.128	-.045	.240
		.030	1.717	1.233	2.135	.235	.091	.329
		.050	1.956	1.454	2.381	.291	.163	.377
		.070	2.147	1.636	2.576	.332	.214	.411
		.100	2.390	1.872	2.820	.378	.272	.450
		.150	2.737	2.217	3.166	.437	.346	.501
		.200	3.048	2.533	3.476	.484	.404	.541
		.250	3.342	2.836	3.771	.524	.453	.576
		.400	4.218	3.734	4.671	.625	.572	.669
		.500	4.851	4.363	5.366	.686	.640	.730
		.600	5.580	5.048	6.227	.747	.703	.794
		.750	7.041	6.302	8.135	.848	.800	.910
		.900	9.845	8.466	12.262	.993	.928	1.089
		.980	15.082	12.135	20.985	1.178	1.084	1.322
		.990	17.532	13.757	25.410	1.244	1.139	1.405

1. Logarithm base = 10

The column labeled *Prob* is the proportion responding. The column labeled *Estimate* is the estimated dosage required to achieve this proportion. For example, you predict that 25% of insects are killed by a dose of 3.34. The 95% confidence limits for the dose are shown in the *Lower Bound* and *Upper Bound* columns. The second part of the table has corresponding values for the log dosages. Look at the probability value of 0.50 to find the previously calculated estimates for the median lethal dose. (The table has been edited to show only some of the probabilities.)

If the chi-square goodness-of-fit test has a significance level of less than 0.15 (the program default), a heterogeneity correction is automatically included in the computation of the intervals (Finney, 1971).

Comparing Several Groups

In the previous example, you have only one stimulus at several doses. If you want to compare several different stimuli, each measured at several doses, additional statistics are useful. Let's add two more insecticides to the previously described problem. Besides rotenone at five concentrations, you also have five concentrations of deguelin and four concentrations of a mixture of the two. Figure 5-6 shows a partial listing of these data, as entered in the Data Editor. As in the previous example, the variable *dose* contains the insecticide concentration, *total* contains the total number of cases, and *died* contains the number of deaths. The factor variable *agent* is coded 1 (*rotenone*), 2 (*deguelin*), or 3 (*mixture*).

Figure 5-6
Data for three insecticides

	dose	agent	total	died
1	2.57	1	50	6
2	3.80	1	48	16
3	5.13	1	46	24
4	7.76	1	49	42
5	10.23	1	50	44
6	10.00	2	48	18
7	20.42	2	48	34
8	30.20	2	49	47
9	40.74	2	50	47
10	50.12	2	48	48
11	10.00	3	46	27
12	15.14	3	48	38
13	20.42	3	46	43
14	25.12	3	50	48

Specifying the Analysis

To analyze the data, open the file *insecticide3.sav* and from the menus choose:

Analyze
  Regression
    Probit...

  ▶ Response Frequency: died
  ▶ Total Observed: total
  ▶ Factor: agent(1,3)
  ▶ Covariate(s) dose

  Transform: Log base 10
  Model
    ⊙Probit

  Options...
    ☑ Relative median potency
    ☑ Parallelism test
    ☑ Fiducial confidence intervals

Fitting the Model

Figure 5-7 is a plot of the observed probits against the logs of the concentrations for each of the three groups separately.

Figure 5-7
Plots of observed probits against logs of concentrations

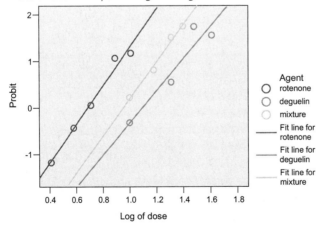

You can see that there appears to be a linear relationship between the two variables for all three groups. The probit model assumes that all three lines are parallel and estimates a common slope for them. Figure 5-8 shows the estimate of the common slope and separate intercept estimates for each of the groups. Based on the Pearson goodness-of-fit test shown in Figure 5-9, the model appears to fit reasonably well.

Figure 5-8
Slope and intercept estimates

			Estimate	Std. Error	Z	Sig.	95% Confidence Interval Lower Bound	Upper Bound
PROBIT[1]	Parameter	dose	3.906	.307	12.728	.000	3.305	4.508
	Intercept[2]	rotenone	-2.673	.236	-11.339	.000	-2.909	-2.438
		deguelin	-4.366	.407	-10.721	.000	-4.773	-3.959
		mixture	-3.712	.375	-9.900	.000	-4.086	-3.337

1. PROBIT model: PROBIT(p) = Intercept + BX (Covariates X are transformed using the base 10.000 logarithm.)
2. Corresponds to the grouping variable agent.

To test whether the three lines are parallel, you can use the parallelism test, which is also shown in Figure 5-9. The observed significance level for the test of parallelism is large—0.559—so there is no reason to reject the hypothesis that all three lines are parallel. The equation for rotenone is estimated to be:

$$\text{Probit}(P_i) = -2.67 + 3.91(\log_{10}(\text{dose}_i))$$

The equation for deguelin is:

$$\text{Probit}(P_i) = -4.37 + 3.91(\log_{10}(\text{dose}_i))$$

and the equation for the mixture is:

$$\text{Probit}(P_i) = -3.71 + 3.91(\log_{10}(\text{dose}_i))$$

Figure 5-9
Chi-square tests

		Chi-Square	df[1]	Sig.
PROBIT	Pearson Goodness-of-Fit Test	7.471	10	.680[2]
	Parallelism Test	1.162	2	.559

1. Statistics based on individual cases differ from statistics based on aggregated cases.
2. Since the significance level is greater than .150, no heterogeneity facto is used in the calculation of confidence limits.

Comparing Relative Potencies of the Agents

The relative potency of two stimuli is defined as the ratio of two doses that are equally effective. For example, the relative median potency is the ratio of two doses that achieve a response rate of 50%. In the case of parallel regression lines, there is a constant relative potency at all levels of response. For example, consider Figure 5-10, which shows some of the doses needed to achieve a particular response for each of the three agents.

Figure 5-10
Confidence limits for effective doses

					95% Confidence Limits for dose			95% Confidence Limits for log(dose)[1]		
					Estimate	Lower Bound	Upper Bound	Estimate	Lower Bound	Upper Bound
PROBIT	agent	rotenone	Prob	.100	2.271	1.885	2.631	.356	.275	.420
				.250	3.249	2.826	3.654	.512	.451	.563
				.500	4.835	4.337	5.376	.684	.637	.730
				.750	7.195	6.435	8.180	.857	.809	.913
				.900	10.291	8.974	12.209	1.012	.953	1.087
		deguelin	Prob	.100	6.159	4.812	7.467	.790	.682	.873
				.250	8.809	7.241	10.328	.945	.860	1.014
				.500	13.110	11.266	14.988	1.118	1.052	1.176
				.750	19.510	17.133	22.253	1.290	1.234	1.347
				.900	27.904	24.381	32.548	1.446	1.387	1.513
		mixture	Prob	.100	4.189	3.203	5.157	.622	.506	.712
				.250	5.990	4.828	7.121	.777	.684	.853
				.500	8.915	7.104	5.855	8.325	.851	.768
				.750	13.268	11.557	15.164	1.123	1.063	1.181
				.900	18.976	16.579	22.002	1.278	1.220	1.342

1. Logarithm base = 10

For rotenone, the expected dosage to kill half of the insects is 4.83; for deguelin, it is 13.11; and for the mixture, it is 8.91. The less of an agent it takes to kill half of the insects, the more potent the agent. The relative median potency for rotenone compared to deguelin is 4.83/13.11 , or 0.37; for rotenone compared to the mixture, it is 0.54; and for deguelin compared to the mixture, it is 1.47. These relative median potencies and their confidence intervals are shown in Figure 5-11. The first estimate in the table is the relative median potency for *agent 1* (rotenone) compared to *agent 2* (deguelin). All pairs of agents are compared.

Figure 5-11
Estimates of relative median potency

					95% Confidence Limits			95% Confidence Limits with LOG Transform [1]		
					Estimate	Lower Bound	Upper Bound	Estimate	Lower Bound	Upper Bound
PROBIT	(I) agent	1	(J) agent	2	.369	.234	.521	-.433	-.632	-.283
				3	.542	.381	.712	-.266	-.419	-.147
		2	(J) agent	1	2.712	1.920	4.282	.433	.283	.632
				3	1.471	1.206	1.850	.167	.081	.267
		3	(J) agent	2	.680	.541	.829	-.167	-.267	-.081
				1	1.844	1.404	2.626	.266	.147	.419

1. Logarithm base = 10

If a confidence interval does not include the value of 1, you have reason to suspect the hypothesis that the two agents are equally potent.

Estimating the Natural Response Rate

In some situations, the response of interest is expected to occur even if the stimulus is not present. For example, if the organism of interest has a very short life span, you would expect to observe deaths even without the agent. In such situations, you must adjust the observed proportions to reflect deaths due to the agent alone.

If the natural response rate is known, it can be entered into the SPSS Probit Analysis procedure. It can also be estimated from the data, provided that data for a dose of 0 are entered together with the other doses. If the natural response rate is estimated from the data, an additional degree of freedom must be subtracted from the chi-square goodness-of-fit degrees of freedom.

More than One Stimulus Variable

If several stimuli are evaluated simultaneously, an additional term is added to the regression model for each stimulus. Regression coefficients and standard errors are displayed for each stimulus. In the case of several stimuli, relative potencies and confidence intervals for the doses needed to achieve a particular response cannot be calculated in the usual fashion, since you need to consider various combinations of the levels of the stimuli.

Kaplan-Meier Survival Analysis

How long do marriages last? How long do people work for a company? How long do patients with a particular cancer live? To answer these questions, you must evaluate the interval between two events—marriage and divorce, hiring and departure, diagnosis and death. Solution of the problem is complicated by the fact that the event of interest (divorce, termination, or death) may not occur for all people during the period in which they are observed, and the actual period of observation may not be the same for all people. That is, not everyone gets divorced or quits a job, and not everyone gets married or starts a job on the same day.

These complicating factors eliminate the possibility of doing something simple, such as calculating the average time between the two events. In this chapter, you'll examine the distribution of time to an event using a method proposed by Kaplan and Meier (1958). It is one of several statistical techniques for analyzing the interval between two events when information is incomplete. Since these techniques are often used to analyze data in which the event of interest is death, they are known as survival-time or failure-time methods. In the social sciences, these methods are often referred to as event history analysis.

SPSS Procedures for Survival Data

SPSS has three procedures for analyzing time-to-event data: Life Tables, Kaplan-Meier, and Cox Regression with and without time-dependent covariates. The survival times are not assumed to come from a particular distribution.

■ The Kaplan-Meier, or product-limit method, estimates the cumulative survival function at the time each event occurs. A row of output is generated for each case in the data file, making results unwieldy for large numbers of cases. The SPSS Kaplan-Meier procedure provides three tests for comparing survival curves based on the values of a single factor. Within each factor, cases can be further subdivided into strata, and comparisons can be done for each factor within a strata or pooled across strata.

- The Life Tables procedure estimates survival at fixed time points after the initial event. It can be thought of as a grouped data version of the Kaplan-Meier procedure, since survival is estimated for cases that are grouped into intervals based on how long they have been observed. Life table output is compact and particularly useful when you have large sample sizes. As the length of the interval decreases, life table estimates approach those from the Kaplan-Meier procedure. You can compare survival curves based on the values of a single factor using the Gehan (or generalized Wilcoxon) test.

- The Cox Regression model, also known as the proportional hazards model, is similar to ordinary regression models in that you predict a dependent variable (length of time to the occurrence of an event) as a function of a set of independent variables. However, unlike ordinary regression models, Cox regression models can be used when there are observations for whom the event has not occurred. If you want to evaluate the effect of one or more covariates on time to an event, this is usually the procedure of choice. If covariates vary with time—for example, if blood pressure is measured three times within a study—you can use a special version of the Cox Regression procedure that incorporates time-dependent covariates.

Background

When you study the time interval between two events, you are interested in the percentage of cases that experience the event at various points in time. This is more informative than calculating a single summary value, such as the median survival time. For example, you want to know what percentage of new car buyers still have the car at various intervals after purchase. You want to say that 80% of buyers still have the car after 1 year; 65%, after 2 years; and 10%, after 10 years. That's easy to do if you observe everyone in the study until they replace the car under study. You simply count the number of consumers who have the same car one year after purchase and divide by the total number of consumers, resulting in a one-year same car "survival" rate. You can easily repeat the calculation for all time points of interest without resorting to elaborate statistical techniques.

The problem becomes more complicated if you don't follow everyone until they part with their car. People who disposed of their car during the study period are called **uncensored** observations, since they experienced the event of interest (parting with their car). People who still have the car in question when your study ends are called **censored** observations (you don't know how long they will keep their car; all you know is that they kept it at least as long as your records indicate). The observations are also called **right-censored**, since you have incomplete information for values in the right tail of the survival distribution.

If you look only at people who purchased cars during the last five years, you have different amounts of information. For people who bought during the last year, you have a year or less of information. For people who bought two years ago, you have two years or less of information. For people who bought five years ago, you have five years or less of information.

How would you go about computing a five-year survival rate? The most straightforward approach is to include only people who bought a car at least five years ago and calculate what percentage still have the car at the five-year mark. If you do that, you're throwing away all of the information that you have for people who have been observed for less than five years. You're wasting good data! That's why you need to analyze your data using specialized survival techniques.

If you analyze your car data with the Kaplan-Meier or Life Tables procedure, you can use all of the data you've gathered to estimate the same-owner survival experience. The idea is simple. You don't deal with actual dates for start and finish times, since they differ; instead, you look at the length of time between the initial event and the terminal event or date of last contact for people who didn't experience the event. Then you subdivide the entire time interval into small pieces and let the data contribute to as many pieces as they can. For example, the probability of surviving three years can be subdivided into the probability of surviving the first year, the probability of surviving the second year, given survival up to the end of the first year, and the probability of surviving the third year, given survival to the end of the second year. All cases that are observed for at least one year contribute to the estimation of the first-year survival probability. Cases that survive the first year contribute to the estimation of the conditional probability of surviving the second year, and so on.

The Life Tables procedure subdivides the time period into equal intervals of your choosing. The Kaplan-Meier procedure creates intervals whose length depends on when events occur. For small interval widths, the life table estimates approach those obtained from the Kaplan-Meier procedure.

Requirements

To use Kaplan-Meier survival analysis, your data must meet certain criteria:

- Observations that are censored must not differ from observations that are not censored. That is, you expect observations for which you have incomplete data to have the same survival experience as observations for which you have complete data. For example, you assume that the overall car retention experience is the same for people who buy cars this year and for people who bought cars five years ago. If that's not true, it makes no sense to pool all of the data and estimate a single set of rates.

- There must be a clearly defined initial event. For example, if you are studying patient survival, you have several choices for the initial event: the date of the first symptoms, the date of the diagnosis, or the date of the first treatment.

- There must be a clearly defined terminal event. "Stone dead hath no equal" as a terminal event. Other end points in a study are often not as well-defined. If you are studying time to divorce, you have several possible terminal events: the date the couple separates, the date the divorce petition is filed, or the date the divorce is granted. Determining whether the terminal event has occurred can also be tricky. For example, what constitutes "relapse," and who decides?

- For uncensored cases—cases that experience the event of interest—you must know the date of the event (or the length of time from the initial event).

- For censored cases—cases for whom the terminal event did not occur—you must know the last date (or the length of time from the initial event) when you observed them to be event free.

Calculating Length of Time

Your data file can contain the actual time (in units of years, days, seconds, or whatever makes sense for your data), or it may contain two dates from which you must calculate the elapsed time. You can calculate elapsed time from the initial event to the terminal event, or last contact, in several ways, depending on the type of data you have in your file.

If you have three variables for each date—*Month, Day,* and *Year*—from the menus choose:

Transform
    Compute...

Use the yrmoda function to calculate the time interval in days between the two events. For example,

Durationdays = YRMODA(*year2*, *month2*, *day2*) − YRMODA(*year1*, *month1*, *day1*)

where *year1*, *month1*, and *day1* are for the first event. Don't forget to convert the results in days into the units you want to use for analysis, such as months or years.

If you have dates in SPSS Date format, you can use the Date and Time Wizard to compute the time interval between two events. To open the Date and Time Wizard, from the menus choose:

Transform
  Date/Time...

For detailed information on using the Date and Time Wizard, see the SPSS online Help system.

Estimating the Survival Function

Consider how a Kaplan-Meier survival curve is calculated. The first column of Table 6-1 contains 10 hypothetical survival times, measured in months, for patients diagnosed with a particular disease. You see that the observed survival times range from 2 months to 72 months. Three of the times are censored. That is, three patients with survival times of 12 months, 40 months, and 69 months were alive at last contact.

Each of the times that corresponds to an observed event defines an interval that begins at the observed event time and ends just before the next ordered event time. (The minus sign at the end of the interval indicates that the end point is not included.) The fourth column tells you the total number of cases that are alive before the current interval. The next column tells you the number of cases that are alive after the current interval.

Estimating the Conditional Probability of Survival

You must estimate a set of conditional probabilities of surviving each interval. This is the probability that a case that enters the interval alive leaves it alive. To calculate this probability, divide the number of cases alive after each event by the number of cases alive in the interval before the event. For example, the probability of surviving at two months is estimated to be 0.90, since 10 cases are alive for up to two months, and only 9 survive past two months. This is shown in the column labeled *Proportion Remaining Alive* in Table 6-1.

Table 6-1
Calculating a Kaplan-Meier curve

Time	Status	Interval	Number Alive Prior	Number Remaining Alive After	Proportion Remaining Alive	Cumulative Survival Proportion
0		0 to 2-	10	10	1	1
2	dead	2 to 5-	10	9	9/10	9/10=0.90
5	dead	5 to 12-	9	8	8/9	0.9x(8/9)=0.80
12	alive	12 to 35-	8	7		0.80
35	dead	35 to 40-	7	6	6/7	0.8x(6/7)=0.69
40	alive	40 to 43-	6	5		0.69
43	dead	43 to 49-	5	4	4/5	0.69x(4/5)=0.55
49	dead	49 to 64-	4	3	3/4	0.55x(3/4)=0.41
64	dead	64 to 69-	3	2	2/3	0.41x(2/3)=0.27
69	alive	69 to 72-	2	1		0.27
72	dead	72+	1	0	0/1	0.27x(0/1)=0

The second death is observed at five months. The probability of surviving to at least five months, assuming that you were alive at two months, is 8 out of 9, since nine cases are alive just prior to five months and only eight cases are alive after five months.

Estimating the Cumulative Probability of Survival

The **cumulative survival probability** estimates the proportion of all cases that are still alive at a particular time point. A cumulative survival of 50% at five months tells you that 50% of the cases are event free five months after the study began. (An interval probability of survival of 50% tells you that half of the cases that are alive at the end of the previous interval are dead by the end of the current interval. It is a conditional probability based on survival to the beginning of the interval.)

To calculate cumulative survival probabilities for a particular event time, you multiply all of the interval survival probabilities up to and including that time. Probabilities are not estimated at the censored times because the proportion surviving does not change at these points. The survival probabilities remain the same as for the preceding noncensored observation.

For example, the cumulative survival at 35 months is:

S(35 months) = 1 x 9/10 x 8/9 x 6/7 = 0.69

If the last observed survival time is associated with an event, the estimated survival probability is 0. That does not mean that all cases died—only that the patient with the longest observed survival time died.

Specifying the Analysis

To compute a Kaplan-Meier table in SPSS, open the file *survival.sav*, and from the menus choose:

Analyze
  Survival
    Kaplan-Meier...

▶ Time: time
▶ Status: dead(1)

Define Event...
⊙ Single value: 1

Options...
  Statistics:
  ☑ Mean and median survival
  ☑ Survival table
  Plots:
  ☑ Survival

The SPSS Kaplan-Meier Table

Figure 6-1 contains output from the SPSS Kaplan-Meier procedure for the previous example. The format is very similar to that of Table 6-1.

Figure 6-1
Kaplan-Meier table

	Time	Status	Cumulative Proportion Surviving at the Time		N of Cumulative Events	N of Remaining Cases
			Estimate	Std. Error		
1	2.000	Dead	.900	.095	1	9
2	5.000	Dead	.800	.126	2	8
3	12.000	Alive	.	.	2	7
4	35.000	Dead	.686	.151	3	9
5	40.000	Alive	.	.	3	5
6	43.000	Dead	.549	.172	4	4
7	49.000	Dead	.411	.176	5	3
8	64.000	Dead	.274	.162	6	2
9	69.000	Alive	.	.	6	1
10	72.000	Dead	.000	.000	7	0

The standard error of the cumulative proportion surviving at time k is estimated using Greenwood's formula:

$$se(S(t_k)) = S(t_k)(\ )\sqrt{\sum_{i=1}^{k}\frac{d_i}{n_i(n_i-d_i)}}$$

where $S(t_k)$ is the cumulative survival probability, d_i is the number of events at time t_i, and n_i is the number of cases alive prior to time t_i. The standard error changes only at event times. As with all nonparametric procedures, for a fixed sample size this results in larger variances, especially for large survival times, than if you assumed a particular survival distribution.

The standard error can be used to construct confidence intervals for the survival curve. However, the confidence interval is not constrained to fall in the range of 0 to 1, and for small sample sizes, the assumption of normality may be tenuous. Other procedures have been suggested (Borgan and Leistol, 1990). If you compute confidence intervals for several time points without adjusting for the fact that you are making multiple comparisons, your joint confidence level will not be correct.

The column labeled *N of Cumulative Events* is a count of all events that have occurred up to and including the current time. The *N of Remaining Cases* column contains the number of cases alive after the current time.

Mean and Median Survival Times

Estimates of the mean and median survival times, their standard errors, and 95% confidence intervals are shown in Figure 6-2. The median survival time is the first observed time at which the cumulative survival is 50% or less. If the cumulative survival distribution doesn't go below 50% for the range of times in the data, the median can't be estimated and doesn't appear in the output. The 95% confidence interval for the median extends from 31 months to 67 months, a fairly wide range given that the largest observed survival time is 72 months. Values anywhere in the range are plausible for the population value of the median survival time.

The mean is not as important a descriptive statistic for survival time as it is for other variables if assumptions are not made about the distribution of the survival times. The median is usually preferred. The mean survival time is *not* the average of the observed survival times, since it does not make sense to compute the usual arithmetic average if all of the cases are not dead. The mean is estimated as the area under the cumulative survival curve. If the largest observed survival time is for a censored observation, the estimate of the mean survival time is said to be restricted to the largest observed survival time.

Figure 6-2

Mean and median survival times

		Mean[1]	Median
Estimate		45.843	49.000
Std. Error		8.728	8.962
95% Confidence Interval	Lower Bound	28.736	31.434
	Upper Bound	62.950	66.566

[1]. Estimation is limited to the largest survival time if it is censored

Plotting Survival Functions

A plot of the cumulative survival function for the sample data is shown in Figure 6-3. Notice that the survival function has steps. That is, it is only at each of the observed event times that the estimate of the survival function changes. It is constant for all times between two adjacent event times. Don't try to connect the points with curves or straight lines, since they distort the true information. A plus sign (+) indicates a censored case.

Factors other than survival times can affect the shape of the survival curve. For example, the distribution of the number of censored cases over the study time changes the appearance of the survival curve.

Figure 6-3
Cumulative survival function

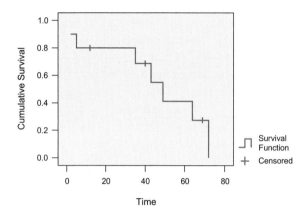

Comparing Survival Functions

In many situations, you want to do more than just estimate a cumulative survival function for a single group. You want to compare survival functions across several groups. Consider the following data from a Veteran's Administration Lung Cancer Trial, presented by Kalbfleisch and Prentice (1980). Survival times, as well as other information, were obtained for a sample of 137 men with advanced inoperable lung cancer.

One of the hypotheses of interest is whether the different histological types of lung cancer affect survival. The patients are divided into four groups based on the histology of their tumors: squamous cell (group 1), small cell (group 2), adenocarcinoma (group 3), or large cell (group 4). Using the SPSS Kaplan-Meier procedure, you can estimate four survival functions, one for each of the histology categories.

Specifying the Analysis

To compare the survival experience of the four groups, open the file *lung.sav*, and from the menus choose:

Analyze
  Survival
    Kaplan-Meier...

▶ Time: time
▶ Status: death(1)
Define Event...
    ⊙ Single value: 1
▶ Factor: histology

Options...
  Plots: Survival

Compare Factor...
  Test Statistics
  ☑ Log rank ☑ Breslow ☑ Tarone-Ware
  ⊙ Pooled over strata

The variable *histology* is specified as a factor variable. Separate survival tables, plots, and comparisons are generated based on the values of the factor variable. You must select Pooled over strata in the Test Statistics group to compare the survival distributions for the factor levels.

Comparing Groups

Figure 6-4 is a summary table of the number of cases and events in each of the histology groups. There is a total of nine censored cases. Eleven percent of the cases in the squamous histology group are censored, 6% in the small-cell group are censored, and almost 4% in both the adenocarcinoma and large-cell groups are censored. It is important to examine the censoring pattern in the groups, since the tests of equality of survival functions depend on the groups having similar patterns. If the pattern of censoring varies markedly among the groups, the test statistic may cause you to draw the wrong conclusion.

Figure 6-4
Summary table for histology factor

			Censored	
Histology	Total N	N of Events	N	Percent
squamous	35	31	4	11.4%
small cell	48	45	3	6.3%
adeno	27	26	1	3.7%
large cell	27	26	1	3.7%
Overall	137	128	9	6.6%

The plot of the estimated survival functions for each of the histology groups is shown in Figure 6-5. From the plot, it appears that the four groups differ in survival. Patients with adenocarcinoma seem to survive the shortest time, while patients with squamous-cell histology appear to have the best prognosis.

Figure 6-5
Survival functions for histology groups

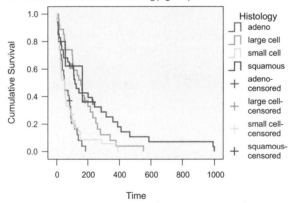

Testing Equality of Survival Functions

Several statistical tests are available for evaluating the null hypothesis that in the population, two or more survival functions are equal. The SPSS Kaplan-Meier procedure lets you choose among three. All three tests are based on computing the weighted difference between the observed and expected number of deaths at each of the time points. All of the tests have little power to detect survival functions that cross (that is, survival functions where one curve is not always above or below another).

A component of the statistic can be written as

$$U = \sum_{i=1}^{k} w_i (0_i - E_i)$$

where w_i is the weight for time point i, and k is the number of distinct time points.

- For the **log rank test**, all of the weights are equal to 1, so the log rank test weights all deaths equally. It is more likely to detect later differences than the other tests.

- For the **Breslow test** (also known as the **generalized Wilcoxon test** and the **Gehan test**), the weights are the number at risk at each time point. The Breslow test weights early deaths more than later deaths because the number at risk always decreases with time, so it is more likely to detect early differences between groups.

- For the **Tarone-Ware test**, the weights are the square root of the number at risk. That's an intermediate strategy.

The log rank test is more powerful than the Breslow test for detecting differences if the mortality rate in one group is a multiple of that in another group. If this is not the case, the log rank test may not be as powerful as the Breslow test. If the percentage of censored cases is large, Prentice and Marek (1979) have shown that the Breslow test has very low power, since the early deaths dominate the statistic.

Figure 6-6 shows the log rank test, as well as the other tests, for testing the null hypothesis that the survival functions are the same for the four histologic groups. Since the observed significance level is small $(p < 0.0005)$, you can reject the null hypothesis. If the three tests lead you to different conclusions, you should be concerned. In particular, make sure to examine the pattern of censoring.

Figure 6-6

Overall test statistics for histology factor

	Chi-Square	df	Sig.
Log Rank (Mantel-Cox)	25.404	3	.000
Breslow (Generalized Wilcoxon)	19.433	3	.000
Tarone-Ware	22.573	3	.000

Test of equality of survival distributions for the different levels of histology.

Based on Figure 6-6, you can reject the overall hypothesis that the four groups have the same survival function, but you can't tell which of the groups are significantly different from each other. To do that, you can compare all pairs of groups.

116

Chapter 6

Recall the Kaplan-Meier dialog box, and in the Test Statistics group, select Pairwise over strata and Log rank.

Figure 6-7 contains the output for all pairwise comparisons. You see that all pairwise comparisons are significant except squamous versus large cell (group 1 versus group 4), and small cell versus adenocarcinoma (group 2 versus group 3).

Figure 6-7

Pairwise log rank tests for histology factor

histology			squamous	small cell	adeno	large cell
Log Rank (Mantel-Cox)	squamous	Chi-Square		11.574	12.045	.823
		Sig.		.001	.001	.364
	small cell	Chi-Square	11.574		.097	9.371
		Sig.	.001		.756	.002
	adeno	Chi-Square	12.045	.097		17.669
		Sig.	.001	.756		.000
	large cell	Chi-Square	.823	9.371	17.669	
		Sig.	.364	.002	.000	

Making all pairwise comparisons results in the problem of multiple comparisons that you've encountered in other procedures, such as one-way analysis of variance. As the number of comparisons increases, the probability of calling a difference significant by chance increases as well. You can protect yourself from rejecting the null hypothesis too often by applying a multiple comparisons procedure.

Stratified Comparisons of Survival Functions

The goal of the Veteran's Administration Lung Trial was to study the efficacy of a new treatment in prolonging survival time of inoperable lung cancer patients. All of the patients were assigned to either the standard treatment or the new treatment. In analyzing the data, your first impulse might be to estimate two survival functions—one for the standard treatment and one for the new treatment—and then to compare them. The problem with this approach is that you have already seen that the type of histology affects the survival prognosis. If the distribution of histology types differs for the two treatments, the histology distribution will affect the treatment comparison. Therefore, you must compare the two treatments within each of the four histology categories.

Specifying the Analysis

When you compare the levels of a factor controlling for the values of another variable, the variable whose effect you want to control for is called the stratification variable. In this example you want to compare the two treatments, or factors, stratifying by histology. You can do this in several ways. You can look at the comparison within each individual stratum. That is, you examine the two treatments within each of the histology types. In SPSS, this is called "for each stratum." You can also pool the results over all of the strata. The comparisons are done for each stratum, but the final result is combined over all of the strata, and you get a single test statistic. In SPSS, this is called "pooled over strata." If you have small numbers of cases in each of the strata, comparisons of the treatments in each stratum will be hampered by small sample sizes.

To compare the two treatments pooling results over strata, recall the Kaplan-Meier dialog box and select:

▶ Time: time
▶ Status: death(1)
Define Event...
   ⊙ Single value: 1
▶ Factor: treatment
▶ Strata: histology

Options...
   Plots
   ☑ Survival

Save...
   ☑ Survival

Compare Factor...
   Test Statistics
   ☑ Log rank ☑ Breslow ☑ Tarone-Ware
   ⊙ Pooled over strata

Distribution of Cases

Figure 6-8 shows the summary table for the data when the cases are classified into both histology groups and treatment categories. From this table, you see that the number of cases receiving the two treatments is not the same in each of the histology strata. For example, of the 48 cases with small-cell carcinoma, 18 were assigned to the new treatment and 30 to the standard treatment.

Figure 6-8
Summary table for histology group and treatment method

histology	treatment	Total N	N of Events	Censored N	Censored Percent
squamous	new	20	18	2	10.0%
	standard	15	13	2	13.3%
	Overall	35	31	4	11.4%
small cell	new	18	17	1	5.6%
	standard	30	28	2	6.7%
	Overall	48	45	3	6.3%
adeno	new	18	17	1	5.6%
	standard	9	9	0	.0%
	Overall	27	26	1	3.7%
large cell	new	12	12	0	.0%
	standard	15	14	1	6.7%
	Overall	27	26	1	3.7%
Overall	Overall	137	128	9	6.6%

Survival Function Plots

To plot the survival functions, from the menus choose:

Graphs
  Scatter/Dot...

Simple Scatter
▶ Y Axis: SUR_1
▶ X Axis: time
▶ Set Markers by: treatment
▶ Panel by: Rows: histology

To connect the points with a step function, double-click the chart to activate it in the Chart Editor. Select a plot marker in the chart, and from the Chart Editor menus choose:

Elements
  Interpolation Line

  ⊙ Step Left step

The survival functions for the two types of treatments for each of the four histology types are shown in Figure 6-9. Within each histology category, you see survival functions for each of the two treatments. From these plots, the cumulative survival doesn't appear to be very different for the two treatments.

Figure 6-9

Plots of survival functions

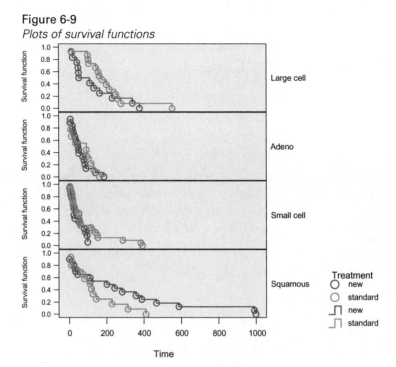

Time

Stratified Overall Comparison of Treatments

You can test the null hypothesis that the two survival functions are the same by pooling the results across the histology category or by evaluating the results separately within each stratum. All three of the previously described tests can be used when the cases are subdivided into strata.

Figure 6-10 shows the results from the three tests when results are pooled over strata. The large observed significance level ($p = 0.402$) tells you that there is not enough evidence to reject the null hypothesis that the survival functions for the two treatments are identical.

Figure 6-10

Test statistics for treatment method

	Chi-Square	df	Sig.
Log Rank (Mantel-Cox)	.702	1	.402
Breslow (Generalized Wilcoxon)	1.044	1	.307
Tarone-Ware	1.023	1	.312

Test of equality of survival distributions for the different levels of treatment.

To obtain results for each individual stratum, recall the Kaplan-Meier dialog box and select For each stratum in the Compare Factor Levels subdialog box. In Figure 6-11, a separate test statistic is shown for each histology category. Again, you can't reject the null hypothesis that there is no difference between the old and new treatments.

Figure 6-11

Comparisons within strata

histology		Chi-Square	df	Sig.
squamous	Log Rank (Mantel-Cox)	2.454	1	.117
small cell	Log Rank (Mantel-Cox)	2.281	1	.131
adeno	Log Rank (Mantel-Cox)	.233	1	.629
large cell	Log Rank (Mantel-Cox)	1.127	1	.288

Test of equality of survival distributions for the different levels of treatment.

Pairwise Comparison of Treatments

In this example, there are only two treatments, so if you reject the null hypothesis that all treatments have the same survival distribution, you don't have to worry about which treatments are different from each other. If you have more than two treatments, you can compare all pairs of treatments, either pooled over all strata or within each of the strata. Remember that sample sizes rapidly diminish as you subdivide your cases into strata and treatments within strata. So interpret the results with caution, especially if you fail to reject the null hypothesis, since you may have limited power for the comparisons of interest.

Figure 6-12 shows the output if you request pairwise comparisons within strata. Since there are only two groups, the results are identical to those in Figure 6-11, although the format is different. For each of the strata, you see a matrix of all possible pairs of factor levels, as well as the chi-square statistic and observed significance level.

If you ask for pairwise tests pooled over strata, there is a single matrix comparing all possible pairs of factor values.

Figure 6-12
Pairwise within strata comparison

	histology	treatment		new	standard
Log Rank (Mantel-Cox)	squamous	new	Chi-Square		2.454
			Sig.		.117
		standard	Chi-Square	2.454	
			Sig.	.117	
	small cell	new	Chi-Square		2.281
			Sig.		.131
		standard	Chi-Square	2.281	
			Sig.	.131	
	adeno	new	Chi-Square		.233
			Sig.		.629
		standard	Chi-Square	.233	
			Sig.	.629	
	large cell	new	Chi-Square		1.127
			Sig.		.288
		standard	Chi-Square	1.127	
			Sig.	.288	

Chapter

7

Life Tables

Many interesting questions can be answered by analyzing the interval between two events. What percentage of people released from prison reenter each year? How likely are you to survive for five years after diagnosis of a serious disease? Are two treatments equally effective in prolonging survival times? When you study the time between two events, you may be faced with two problems: Not everyone in the study experiences the event of interest, and people are observed for differing lengths of time. Not everyone reenters prison. You don't want to wait to publish your results until your last patient dies.

SPSS has three procedures for analyzing length of time to an event data. They are described in Chapter 6. This chapter describes one of the oldest methods for analyzing this type of data—life tables.

Background

Follow-up, or actuarial, life tables were first applied to the analysis of population survival data, from which the term **life table** originates. The basic idea of the life table is to subdivide the period of observation after a starting point, such as beginning work at a company, into smaller time intervals—say, single years. For example, you decompose the probability of working for a company for five years into the probability of working for the company for one year, the probability of working for the company for two years, given that you survived for one year, the probability of working for three years, given that you survived for two years, and so on.

For each interval, all people who have been observed at least that long are used to calculate the probability that a **terminal event**, such as leaving the company, will occur in that interval. Everybody contributes to the probability of surviving one year. People who have survived one year also contribute to the calculation of the probability of surviving for two years, given that you survived for one. A case contributes to the estimation of survival probabilities for all intervals up to, and including, the interval in which the event of interest occurs or the case is lost to observation. The probabilities estimated from each of the intervals are then used to estimate the overall probability of the event occurring at different time points. This makes the best use of the available information.

Studying Employment Longevity

As an example of life table techniques, consider the following problem: As personnel director of a small company, you are asked to prepare a report on the longevity of employees in your company. Information on the date that employees started at the company and the last date they worked is available in the corporate database. You know that it is wrong to look at just the average time on the job for people who left. It doesn't tell you anything about the length of employment for people who are still employed. For example, if your only departures were 10 people who left during their first year with the company, an average employment time based only on them would be highly misleading. You need a way to use information from both the people who left and those who are still with the company.

The employment times for people who are still with the company are known as **censored observations**, since you don't know exactly how long these people will work for the company. You do know, however, that their employment time will be at least as long as the time they have already been at the company. Since you know only the minimum time for people with incomplete data, the observations are said to be **right-censored**. People who have already left the company are **uncensored**, since you know their employment times exactly.

Specifying the Analysis

To obtain a life table, open the file *employment.sav*, and from the menus choose:

Analyze
  Survival
    Life Tables...

▶ Time: length
  Display Time Intervals 0 through 9 by 1

▶ Status: Left(1)

Options...
  ☑ Life table(s)

The Body of a Life Table

The first step in computing a life table is to construct a summary table such as that shown in Figure 7-1. From this table, you can easily calculate all of the life table statistics.

Figure 7-1
Counts of events

Life Table[1]

Interval Start Time	Number Entering Interval	Number Withdrawing during Interval	Number Exposed to Risk	Number of Terminal Events
0	100	2	99.000	2
1	96	2	95.000	1
2	93	16	85.000	7
3	70	15	62.500	6
4	49	12	43.000	5
5	32	10	27.000	5
6	17	9	12.500	4
7	4	1	3.500	1
8	2	2	1.000	0

[1]. The median survival time is 6.43

Interval Start Time is the beginning value for each time interval. The first interval starts at time 0 and goes up to, but does not include, one year; the second interval corresponds to a time period of one year or more but less than two, and so on, for all years up to and including eight. Each interval extends from its start time up to, but not including, the start time of the next interval.

In this example, each row of the table corresponds to a time interval of one year, since that is the "by" value under Display Time Intervals in the Life Tables dialog box, and the data are recorded in years. If you specify 1 when the data are recorded in days, the interval length is one day. (You can choose any interval length you want. For a rapid-turnover company, you might want to consider monthly intervals; for a company with minimal turnover, you might want to consider intervals of several years.)

Number Entering Interval is the number of cases with event times that are at least as long as the interval start time. It is the number of cases that "survive" to the beginning of the current interval. For the first interval, the number is 100, since you have information for 100 people. The next interval, which corresponds to a survival time of at least one year, has 96 people entering it. That means that four people who entered the study failed to survive to one year, either because they quit before they were at the company for one year or because they are still employed but have been with the company for less than one year. You can tell how many cases fall into each of these two categories from two other columns in the table.

Number Withdrawing during Interval is the number of people who worked at the company for less than one year and are still there. It is the number of cases entering the interval for which follow-up ends somewhere in the interval. These are censored cases; that is, these are cases for which the event of interest has not occurred at the time of last contact.

Number Exposed to Risk is the number of people entering the interval minus one-half of those who withdrew during the interval. This is an estimate of how many persons were observed during the entire interval. This is also known as the **effective sample size**. You assume that cases that withdraw during an interval do so at the midpoint of the interval; thus, each contributes only one-half of an observation for that interval. For example, for the first interval, instead of observations for 100 people, you have observations for 99 $(100 - (0.5 \times 2))$.

Number of Terminal Events is the number of cases for which the event of interest occurs within the interval. For the first interval, two people left the company. That is, they experienced the event of interest.

Calculating Survival Probabilities

Based on the counts in Figure 7-1, you can calculate summary statistics that describe the longevity of the employees. The objective is fairly simple. For each interval in the life table, based on people who entered the interval without experiencing the event, you calculate the probability of "surviving" to the end of the interval. You then multiply these probabilities to get cumulative survival estimates at the end of each interval.

The advantage of this approach is that you extract as much information as possible from your data. For example, if you want to calculate what percentage of people are still with your company five years after joining without using a life table approach, you would look only at the records of those who were hired five years ago and count how many are still employed at the end of five years. You would ignore the information you have in your files for employees who were hired less than five years ago. With a life table approach, you use the information from all people. Everybody contributes to the calculation of the survival probabilities. All people who have been around for at least a year contribute to the calculation of the probability of surviving for a year, all people who have been around for at least two years contribute to the calculation of the probability of surviving two years, and so on.

For example, from Figure 7-1, you know that you have 100 employees. Since 2 of them are recent employees who haven't been around for the entire year, you estimate that you have 99 employees' worth of information for the first interval. Since 2 employees quit during the first year, the probability that an employee will survive to the end of their first year is

prob(survive 1 year) = 97/99 = 0.980.

At the beginning of year 2 (which is the end of year 1), there are 96 employees at the company. Two of them are censored, meaning that they have been employed for less than two full years, and one employee leaves. The probability that an employee who survived to the beginning of year 2 will survive to the end of year 2 is (95–1) / 95 = 0.989, where 95 is the number exposed to risk in the interval and 1 is the number of events in the interval.

You can combine these probabilities to calculate the probability that an employee will survive to the end of year 2 using the fact that this probability is the probability that an employee will survive to the end of the first year times the probability that an employee who has survived to the end of the first year will survive to the end of the second year:

P(survival at end of year 2) = P(survival to end of year 1) × P(survival to end of year 2 given survival to end of year 1).

P(survival at end of year 2) = 0.980 × 0.989 = 0.969

You can calculate the probability of surviving to the end of three years by multiplying the P(survival at end of year 2) by the probability of surviving to the end of year 3, given survival to the end of year 2. You can continue calculating successive probabilities in the same way. Of course you don't have to actually do anything, since SPSS will perform the computations for you.

Life Table with Survival Probabilities

A life table with survival probabilities is shown in Figure 7-2, and a description of the columns follows.

Figure 7-2

Life table with survival probabilities

Life Table[1]

Interval Start Time	Number Entering Interval	Number Withdrawing during Interval	Number Exposed to Risk	Number of Terminal Events	Proportion Terminating	Proportion Surviving	Cumulative Proportion Surviving at End of Interval	Std. Error of Cumulative Proportion Surviving at End of Interval
0	100	2	99.000	2	.020	.980	.980	.014
1	96	2	95.000	1	.011	.989	.969	.017
2	93	16	85.000	7	.082	.918	.890	.033
3	70	15	62.500	6	.096	.904	.804	.045
4	49	12	43.000	5	.116	.884	.711	.056
5	32	10	27.000	5	.185	.815	.579	.070
6	17	9	12.500	4	.320	.680	.394	.090
7	4	1	3.500	1	.286	.714	.281	.115
8	2	2	1.000	0	.000	1.000	.281	.115

[1] The median survival time is 6.43

Proportion Terminating is the probability that the event of interest will occur during the interval for a case that has survived to the beginning of that interval. For example, for the interval with a start time of five years, it is the probability that a person who has worked at a company for at least five years will quit before reaching the sixth year. For each interval, it is the number of events divided by the number at risk. For example, the probability that an employee who has been with the company for five years will quit during the sixth year of employment is 5/27, or 0.185, where 5 is the number of terminal events for this interval and 27 is the number at risk.

Proportion Surviving is 1 minus the proportion terminating. It's the probability that a person who enters the interval will leave the interval without the event occurring. For the interval that starts with five years, it is $1 - 0.185 = 0.815$.

Cumulative Proportion Surviving at End of Interval is the estimate of the probability of surviving to the end of the interval. For example, the cumulative proportion "surviving" (still employed) at the end of five years is 0.711. You estimate that almost 71% of employees are still employed at the *end* of five years. Usually, this is the number you are most interested in. You want to know what percentage of patients are still alive 10 years after diagnosis and what percentage of discharged felons are in jail 2 years after discharge.

Std. Error of Cumulative Proportion Surviving at End of Interval is an estimate of the variability of the estimate of the cumulative proportion surviving.

Hazard Rate and Probability Density

There are several additional statistics, shown in Figure 7-3, that are useful when you are analyzing survival data.

Figure 7-3
Hazard rate and probability density function

Life Table[1]

Interval Start Time	Number Entering Interval	Number Exposed to Risk	Number of Terminal Events	Cumulative Proportion Surviving at End of Interval	Probability Density	Std. Error of Probability Density	Hazard Rate	Std. Error of Hazard Rate
0	100	99.000	2	.980	.020	.014	.020	.014
1	96	95.000	1	.969	.010	.010	.011	.011
2	93	85.000	7	.890	.080	.029	.086	.032
3	70	62.500	6	.804	.085	.033	.101	.041
4	49	43.000	5	.711	.094	.040	.123	.055
5	32	27.000	5	.579	.132	.054	.204	.091
6	17	12.500	4	.394	.185	.080	.381	.187
7	4	3.500	1	.281	.113	.099	.333	.329
8	2	1.000	0	.281	.000	.000	.000	.000

[1]. The median survival time is 6.43

Probability Density is an estimate of the probability per unit time of experiencing an event in the interval.

Std. Error of Probability Density is an estimate of the variability of the estimated probability density.

Hazard Rate is an instantaneous rate of change of the death probability at time t, for a person who has survived to time t. It is also known as the force of mortality, or the conditional mortality rate. The hazard multiplied by a small increment of time is approximately the probability that the event will occur in that small increment of time. The hazard rate is modeled in the Cox Regression procedure.

Std. Error of Hazard Rate is an estimate of the variability of the estimated hazard rate.

Median Survival Time

An estimate of the median survival time is displayed in the footnote below the life table shown in Figure 7-2. The median survival time is the time point at which the value of the cumulative survival function is 0.5. That is, it is the time point by which half of the cases experience the event. Linear interpolation is used to calculate this value. If the cumulative proportion surviving at the end of the last interval is greater than 0.5, the start time of the last interval is identified with a plus sign (+) to indicate that the median time exceeds the start value of the last interval.

From Figure 7-2, you see that the median survival time for the personnel data is 6.4 years. This means that half of the employees have left the company after 6.4 years. At the end of nine years, only 28% of the employees remain.

Assumptions Needed to Use the Life Table

The basic assumption underlying life table calculations is that survival experience does not change during the course of the study. For example, if the employment possibilities or work conditions change during the period of the study, it makes no sense to combine all of the cases into a single life table. To use a life table, you must assume that a person hired today will behave in the same way as a person who was hired five years ago. You must also assume that observations that are censored do not differ from those that are not censored. These are critical assumptions that determine whether life table analysis is an appropriate technique.

Lost to Follow-up

In the personnel example, information was available for all employees in the company. We knew who left and when. This is not always the case. If you are studying the survival experience of patients who have undergone a particular therapeutic procedure, there may be two different types of censored observations. There will be patients whose length of survival is not known because they are still alive. There may also be patients with whom you have lost contact and for whom you know only that they were alive at some date in the past. If patients with whom you have lost contact differ from patients who remain in contact, the results of a life table analysis will be misleading.

Consider the situation in which patients with whom you have lost contact are healthier than patients who remain in contact. By assuming that patients who are lost to follow-up behave in the same way as patients who are not, the life table will underestimate the survival experience of the group. Similarly, if sicker patients lose contact, the life table will overestimate the proportion surviving at various time points. It cannot be emphasized too strongly that no statistical procedure can atone for problems associated with incomplete follow-up.

Plotting Survival Functions

The survival functions displayed in the life table can also be plotted. This allows you to better examine the functions. It also allows you to compare the functions for several groups. For example, Figure 7-4 is a plot of the cumulative percentage surviving when the employees from Figure 7-2 are subdivided into two groups: clerical and professional (you specify *type* as the factor variable). You see that clerical employees have shorter employment times than professional employees.

To obtain the plot and a comparison of the survival distribution for the two groups of employees, recall the Life Tables menu and choose:

Analyze
 Survival
  Life Tables...

▶ Time: length
 Display Time Intervals 0 through 9 by 1
▶ Status: left

Define Event...
 ⊙ Single value: 1

▶ Factor: type (1 2)
 Define Range...
 Minimum 1
 Maximum 2

Options...
 Plot
 ☑ Survival

 Compare Levels of First Factor
 ⊙ Overall

Figure 7-4
Plot of survival for hourly and salaried employees

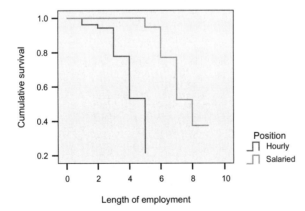

Comparing Survival Functions

If two or more groups in your study can be considered as samples from some larger population, you may want to test the null hypothesis that the survival distributions are the same for the subgroups. The SPSS Life Tables procedure uses the Wilcoxon (Gehan) test (Hosmer and Lemeshow, 1999). The Kaplan-Meier procedure provides additional tests.

In Figure 7-5, the observed significance level for the test that all groups come from the same distribution is less than 0.0005, leading you to reject the null hypothesis that the groups do not differ in the distribution of survival times.

Figure 7-5
Wilcoxon test of equality of survival functions

Overall Comparisons[1]

Wilcoxon (Gehan) Statistic	df	Sig.
26.786	1	.000

[1] Comparisons are exact.

Cox Regression

The time interval between any two interesting events usually depends on factors other than chance alone. For example, the duration of a marriage is influenced by children, finances, and compatibility. Similarly, the survival time after diagnosis of a disease depends on predictor variables such as the severity of the disease, treatment method, and general condition of the patient, among numerous other factors.

In Chapter 6 and Chapter 7, you examined the time interval between two events using Kaplan-Meier and actuarial estimators of the survival distribution. To use these methods to assess the influence of predictor variables on survival times, you must calculate separate survival distributions for each of the categories of a predictor variable. Values of continuous variables, such as age, must be grouped into categories.

Analyses requiring that cases be subdivided into groups based on the values of the predictor variables are often unsatisfactory, since the number of cases in a group rapidly diminishes with increasing numbers of predictor variables. That's why techniques such as multiple linear regression are invaluable for studying the relationship between a dependent variable and a set of predictor variables. However, multiple linear regression cannot be used for analysis of time-to-event data, since there is no way to handle censored observations, or cases for which the event of interest has not yet occurred. In this chapter, you'll use a special type of regression model, the **Cox proportional hazards model**, to analyze data that contain censored observations.

There are two SPSS procedures for proportional hazards regression. The Cox Regression procedure analyzes models with **time-constant covariates** (covariates that do not change as a function of time). The Time-Dependent Cox Regression procedure analyzes models containing one time-dependent covariate, one or more time-constant covariates, or both. Both procedures use the Newton-Raphson method to estimate model parameters. However, you should use the Cox Regression procedure if your model contains only time-constant covariates, since this procedure allows you to save diagnostic variables and obtain plots that are not available in the Time-Dependent Cox Regression procedure.

SPSS currently does not explicitly offer parametric survival modeling, although a simple parametric proportional hazards model can be estimated using the General Loglinear Analysis procedure.

The Cox Regression Model

The regression model proposed by Cox (1972) can be expressed in several equivalent ways. It can be written in terms of the cumulative survival function, the hazard function, and the log of the hazard function. Consider the simple case with only one predictor variable, say *age*. (In the Cox model, the risk, or predictor, variables are usually called **covariates**.)

When the **cumulative survival function**—that is, the proportion of cases "surviving" at a particular point in time—is the dependent variable, the model is

$$S(t) = [S_0(t)]^p$$

where

$$p = e^{(B \times age)}$$

From the model, you see that for any value of the variable *age*, the proportion surviving at time *t* depends on two quantities. The first, designated as $S_0(t)$ in the equation, is called the **baseline survival function**. The baseline survival function does not depend on age; it depends only on time. The baseline survival function is similar to the constant term in multiple regression in that it is the reference value that is increased or decreased depending on the values of the independent variables and their relationship with the dependent variable. The second part of the model, designated as *p*, depends not on time but on the value of the covariate *age* and on the value of the regression coefficient *B*.

The Hazard Function

The cumulative survival function is probably the most intuitive way of characterizing survival times. However, several other closely related functions are also used. The **hazard function (h(t))**, or death rate at time *t*, tells you how likely a case is to experience an event given that the case has survived to that time. The hazard function is not a probability but a death rate per unit of time, so it need not be less than 1. The **cumulative hazard function (H(t))** at time *t* is the sum of all hazard values up to and including those at time *t*. Note that this is not the sum of the hazards for the observed survival times but the mathematically derived sum for all times (the integral).

The cumulative hazard function (H(t)) and the cumulative survival function (S(t)) are related by the following equation:

$$S(t) = e^{-H(t)}$$

The Cox regression model is written in terms of the hazard function as

$$h(t) = [h_0(t)]e^{(BX)}$$

The hazard function, like the survival function, is factored into two component pieces. The baseline hazard, $h_0(t)$, depends only on time, while $e^{(BX)}$ depends only on the values of the covariates and the regression coefficients. Instead of raising the cumulative survival function to the $e^{(BX)}$ power, the baseline hazard is multiplied by $e^{(BX)}$.

If you divide both sides of the equation by $h_0(t)$ and then take logarithms of both sides of the equation, the result is

$$\log\left(\frac{h(t)}{h_0(t)}\right) = BX$$

Continuous variables are assumed to be linearly related to the log of the hazard function. If a regression coefficient is greater than 0, the hazard increases and the cumulative survival decreases. If the coefficient is less than 0, the hazard decreases and the cumulative survival increases.

Since the Cox model for the hazard function results in a simpler equation than that for the survival function, the Cox model is usually expressed in hazard form and is called the **Cox proportional hazards model**. The model is termed *semiparametric* because, although it doesn't make assumptions about the underlying cumulative hazard function, it assumes that covariates are additive and linearly related to the log of the hazard function. For computational details on the method of partial likelihood (the technique used for estimating the model parameters), see Hosmer and Lemeshow (1999) or Klein and Moeschberger (1997).

Proportional Hazards Assumption

The reason for the name "proportional hazards" is the fact that for any two cases, the ratio of their predicted hazards is constant for all time points. For example, the ratio of the hazards for two cases, one with an age of 20 and the other with an age of 40, is

$$HR(t) = e^{B(20-40)}$$

For any value of time, the ratio (not the difference) is constant and does not involve the value of *t*. The assumption of proportional hazards for two or more groups implies that the cumulative survival curves (and the hazard curves) for the groups do not cross. (For more information, see "Checking the Proportional Hazards Assumption" on p. 150.)

The assumption that hazards are proportional restricts the applicability of the Cox regression model. There are situations in which such an assumption is not realistic. For example, if you are comparing surgical and medical treatments, it is quite likely that for early time points, the hazards associated with surgery exceed those of the medical treatment. However, at later time points, the relationship between the two hazard curves may change. In this situation, the ratio of the hazard functions is not a constant over time. Similarly, for a variable such as *age* (at diagnosis), the hazard rate may increase more from age 60 to age 70 than from age 20 to age 30. The proportional hazards assumption dictates that the ratio of the hazards must be the same for all age intervals.

Modeling Survival Times

To better understand the components of the Cox regression model, consider a data set presented by Bartolucci and Fraser (1977). The data track the survival experience of 60 patients with Hodgkin's disease who were receiving standard treatment for the disease. Age, sex, stage of the disease, histology, survival time in months, and censoring status are recorded for each patient.

The first step in analyzing survival data is the same as that in all statistical analyses: visual examination of the data. In the analysis of survival data with censored observations, plots are necessarily more complex because you must consider an additional component whether or not a data point is censored. One way to incorporate the censoring pattern in a display is to choose a different plotting symbol for censored and uncensored points. The failure times are more important, so the uncensored cases should be more prominent.

For example, Figure 8-1 is a plot of survival time against age with censored cases represented by open circles and uncensored cases represented by closed circles. For the very early ages, there appears to be somewhat more censoring. There also appears to be a relationship between age and survival time. The proportional hazard model assumes that there is a linear relationship between the values of age and the log of the hazard function.

Figure 8-1
Survival time against age

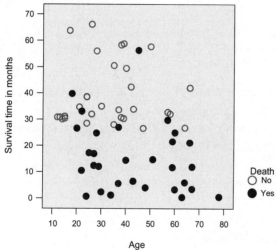

Coding Categorical Variables

If you have categorical covariates, their values must be transformed by creating a new set of variables that correspond in some way to the original categories. The Cox Regression procedure does this automatically when, in the Define Categorical Covariates dialog box, you specify that a variable is categorical. The number of new variables required to represent a categorical variable is one fewer than the number of categories. A separate coefficient is estimated for each new variable.

The interpretation of the coefficients for the coded variables depends on the coding scheme selected. If you include interactions between categorical variables, be sure that you include the main effects as well so that the results are readily interpretable. (For further discussion of contrasts, see the chapter "Logistic Regression Analysis" in the *SPSS 14.0 Statistical Procedures Companion*.)

The available coding schemes for categorical variables are:

- **Indicator (also known as dummy coding)**. This is the default. Each category of the predictor variable except the reference category is compared to the reference category.

- **Simple**. Each category of the predictor variable except the reference category is compared to the reference category. Note that indicator and simple contrasts result in the same parameter estimates except for the constant.

- **Difference**. Also known as reverse Helmert contrasts. Each category of the predictor variable except the first category is compared to the average effect of previous categories.

- **Helmert**. Each category of the predictor variable except the last category is compared to the mean effect of subsequent categories.

- **Repeated**. Each category of the predictor variable except the last category is compared with the category that follows it.

- **Polynomial**. The predictor variable is transformed into linear, quadratic, and cubic components, and so on (depending on the number of categories). Categories are assumed to be equally spaced.

- **Deviation**. Each category of the predictor variable except the reference category is compared to the average of all levels.

For deviation, simple, and indicator contrasts, you can select either First or Last (the default) as the reference category.

Specifying the Analysis

To examine the effect of age upon survival, you can calculate a Cox regression model with survival time as the dependent variable and the age of the patient at diagnosis as the covariate.

Open the file *hodgkins.sav*, and from the menus choose:

Analyze
  Survival
    Cox Regression...

▶ Time: survivaltime

▶ Status: dead
    Define Event...
    Single value: 1

▶ Covariates: age

Options...
  ☑ Display baseline function

Testing Hypotheses about the Coefficient

Figure 8-2 contains statistics for the variable *age* when it is entered into the Cox regression model. The coefficient for *age* is 0.03, with an observed significance level of 0.009, based on the Wald test. You can reject the null hypothesis that the population value of the coefficient is 0. The Wald statistic is the square of the ratio of the coefficient to its standard error. For large sample sizes, under the null hypothesis that the coefficient is 0, the Wald statistic has a chi-square distribution.

Even though you reject the null hypothesis that the population value for the *age* coefficient is 0, it doesn't mean that *age* is linearly related to the hazard function. It simply means that there may be a linear component. (For more information, see "Examining the Functional Form of a Covariate" on p. 156.)

Figure 8-2
Statistics for age

	B	SE	Wald	df	Sig.	Exp(B)
age	.030	.012	6.803	1	.009	1.031

Although you can use the Wald statistic to test hypotheses about the regression coefficients, other statistics are often preferred. Figure 8-3 contains two additional tests that the population value of the coefficient is 0: the likelihood-ratio test and the score test. The likelihood-ratio test is the change in –2 partial log-likelihood when *age* is added to a model. Since *age* is the only variable in this model, it is the change from a model without any variables. The score test is a computationally less intensive statistic based on the partial likelihood. Usually, the values of the three statistics are fairly comparable. If the statistics suggest different conclusions, the likelihood-ratio test is thought to be the most reliable.

Figure 8-3
Omnibus tests of model coefficients

Omnibus Tests of Model Coefficients [1]

-2 Log Likelihood		214.378
Overall (score)	Chi-square	7.104
	df	1
	Sig.	.008
Change From Previous Step	Chi-square	6.844
	df	1
	Sig.	.009
Change From Previous Block	Chi-square	6.844
	df	1
	Sig.	.009

[1] Beginning Block Number 0, initial Log Likelihood function:
-2 Log likelihood: 221.222

Interpreting the Regression Coefficient

As is the case in other regression models, the value of the regression coefficient doesn't indicate the "importance" of a variable. The actual value of the coefficient depends on the units of measurement of the variable, as well as on the other variables in the model. Variables with positive coefficients are associated with decreased survival times (increased hazard), while variables with negative coefficients are associated with increased survival times (decreased hazard).

The column labeled *Exp(B)* in Figure 8-2 is useful for describing the impact of a predictor variable. It is the ratio of the hazard rates for cases that are one unit apart in the values of the predictor variable. For example, a one-year increase in age increases the hazard rate by 3.1%, since the ratio of the hazard rates for cases one year apart in age is 1.031. Different units of measurement for the same variable will change the hazard ratio. You'll get different values if age is in months or in years. Even values close to 1 can be important when meaningful changes in the covariate are considered. For example, a decade of *age* in this example translates to a hazard ratio of 1.031^{10}, which is 1.36. The predicted hazard rate increases by 36% for each decade of increase in *age*.

If the covariate is a dichotomous variable coded so that the two values are one unit apart (for example, 0, 1 or 1, 2) and the larger of the two indicates presence of the characteristic, e^B is the ratio of the estimated hazard for a case with the characteristic to that for a case without the characteristic. This is often called the **relative risk** associated with the variable. If the relative risk is 1, the variable does not influence survival. If the relative risk is less than 1, a positive value for the variable is associated with increased survival, since the hazard rate is decreased. The relative risk can range from 0 to infinity.

Baseline Hazard and Cumulative Survival Rates

From the regression coefficients and the baseline survival function, you can estimate the cumulative survival function for different values of the covariates. (In the usual regression model, you generate one predicted value for each case. In the Cox regression model, you generate an entire survival curve for each combination of values of the predictor variables.)

For example, to calculate the estimated survival curve for 20-year-olds, you use the equation

$$\hat{S}(t|age = 20) = [\hat{S}_0(t)]^{1.82}$$

The exponent is $e^{(0.030 \times 20)}$, in which 0.030 is the regression coefficient for *age* and 20 is the age group of interest. You can estimate the survival curve for persons of any age simply by changing the value of *age* in the previous equation. All of the curves generated in this fashion are just different powers of the baseline survival function.

If the value of the exponent to which the baseline survival curve is raised is greater than 1, the resulting survival times are smaller than those of the baseline. If the exponent to which the baseline survival curve is raised is less than 1, the resulting survival times are greater than those of the baseline.

Figure 8-4 is an excerpt from a table of baseline functions that SPSS displays. There is a row for every observed event time in the data file. The column labeled *Baseline Cum Hazard* is an estimate of $H_0(t)$—the baseline cumulative hazard function. It depends only on time. It is the value when all covariates are 0. You can calculate the estimated baseline cumulative survival function using the relationship

$$S_0(t) = e^{-H_0(t)}$$

If you don't have any covariates, the estimated baseline cumulative survival is the same as that from the Kaplan-Meier procedure.

The columns under *At mean of covariates* are the estimated cumulative survival and the estimated cumulative hazard when all covariates are at the mean, which is shown in Figure 8-5. From this table, your estimate of the cumulative survival rate at 3.5 years, for persons with an average age of 38.7, is 0.894. The estimated cumulative hazard at 3.5 years is 0.112 for persons of the average age.

Figure 8-4
Baseline functions

Time	Baseline Cum Hazard	At mean of covariates		
		Survival	SE	Cum Hazard
.37	.005	.985	.014	.015
.53	.009	.970	.021	.030
.70	.014	.955	.025	.046
1.17	.019	.940	.029	.062
2.37	.024	.925	.033	.078
3.30	.030	.909	.036	.095
3.50	.035	.894	.038	.112

Figure 8-5
Mean of covariates

	Mean
age	38.733

Including Multiple Covariates

Like multiple regression, the Cox regression model can be used for any number of predictor variables. The general form of the model is

$$h(t) = [h_0(t)]e^{(B_1X_1 + B_2X_2 + \ldots + B_pX_p)}$$

where X_1 to X_p are the covariates. The covariates can be either continuous variables, such as age, blood pressure, and temperature, or categorical variables, such as stage of a disease, histology type, or treatment.

Model with Three Covariates

In the Hodgkin's disease example, you found *age* to be significantly associated with the hazard of dying. Consider what happens when two additional variables, *hist* (coded 1 for *nodular sclerosis*, 2 for *mixed cellular*, and 3 for *lymphocyte depletion*) and *stage* (coded 0 for *less advanced* and 1 for *more advanced*), are entered into the model that already contains *age*.

Recall the Cox Regression dialog box and select:

▶ Covariates: age, hist, stage

Categorical...
▶ Categorical Covariates: hist(Indicator), stage(Indicator(first))

Method: Enter

Options...
 Model Statistics
 ☑ CI for exp(B)

Since *stage* is already coded as an indicator variable, there is no need to declare it as categorical unless you want to use it to define separate groups in plots. If you declare it as categorical and specify indicator coding, by default the last category is used as the reference category. This means that the coefficient displayed will be for the less-advanced category. Specify the first category as the reference category to override this.

The coding for the histology variable is shown in Figure 8-6. The last category, lymphocyte depletion, is the reference category to which nodular sclerosis and mixed cellular types are compared.

Figure 8-6
Coding for categorical variables

Categorical Variable Codings[2,3]

		Frequency	(1)	(2)
stage[1]	0=Early	26	0	
	1=Advanced	34	1	
hist[1]	1=nodular sclerosis	32	1	0
	2=mixed cellular	21	0	1
	3=lymphocyte depletion	7	0	0

[1.] Indicator Parameter Coding

[2.] Category variable: stage (stage of disease)

[3.] Category variable: hist (histology)

From the column labeled *B* in Figure 8-7, the estimated model is

$$\hat{h}(t) = [\hat{h}_0(t)]e^{(0.030 \times age + 0.847 \times stage - 1.18 \times hist(1) - 1.32 \times hist(2))}$$

The standard errors of the coefficients are shown in the column labeled *SE*. The square of the ratio of the coefficient to its standard error is in the column labeled *Wald*. For a categorical variable, there is an overall test of the null hypothesis that all coefficients for that effect are 0, as well as parameter estimates and test statistics for each level. The degrees of freedom associated with the statistic are 1 unless the test is for a categorical variable, in which case the degrees of freedom for the overall test of the variable are equal to the number of categories of the variable minus 1. The column labeled *Sig.* shows that the coefficients for all variables are significantly different from 0.

Figure 8-7
Statistics for the three-variable model

	B	SE	Wald	df	Sig.	Exp(B)	95.0% CI for Exp(B)	
							Lower	Upper
age	.030	.012	5.831	1	.016	1.031	1.006	1.056
hist			6.309	2	.043			
hist(1)	-1.179	.545	4.679	1	.031	.307	.106	.895
hist(2)	-1.315	.545	5.829	1	.016	.269	.092	.781
stage	.847	.410	4.262	1	.039	2.334	1.044	5.217

As in linear regression and logistic regression models, the value of the coefficients and their significance levels depends not only on the strength of the association of the individual variables with the dependent variable but also on the other covariates in the model. If some of your independent variables are highly correlated, the resulting coefficients may be misleading. That is, variables may have wrong signs and the significance levels may not correctly reflect the strength of the association with the dependent variable. In such situations, you should include only one of the set of highly correlated variables in the model.

As shown in Figure 8-7, the relative risk for the advanced stage of the disease is 2.33. This means that the estimated risk of dying at any time is 2.33 times greater for a person with advanced disease compared with a person without advanced disease, adjusting for the other factors in the model. You can calculate confidence intervals for e^B by finding the confidence interval for β, the population value, and then raising the lower limit and the upper limit to the power of e. The 95% confidence intervals for e^B are given in the last two columns of Figure 8-7. If the 95% confidence interval does not include the value of 1, you can reject the null hypothesis that the variable is not related to survival.

Global Tests of the Model

Whenever you build a statistical model, you want an overall test of the hypothesis that all parameters are 0. In the Cox regression model, as in logistic regression, several somewhat different statistics can be used. For large sample sizes, they all have a chi-square distribution. Since the estimation method used in the Cox Regression procedure depends on maximizing the partial likelihood function, comparing changes in the values of –2 times the log-likelihood (*–2LL*) for two models that differ only in the inclusion of a subset of variables is the basis for a variety of tests.

The likelihood-ratio test for the hypothesis that all parameters are 0 is obtained by comparing *–2LL* for a model in which all coefficients are 0 (the initial model) with *–2LL* for a model that contains all of the coefficients of interest.

Figure 8-8
Omnibus tests of model coefficients

Omnibus Tests of Model Coefficients [1,2]

-2 Log Likelihood		202.881
Overall (score)	Chi-square	24.980
	df	4
	Sig.	.000
Change From Previous Step	Chi-square	18.340
	df	4
	Sig.	.001
Change From Previous Block	Chi-square	18.340
	df	4
	Sig.	.001

[1] Beginning Block Number 0, initial Log Likelihood function: -2 Log likelihood: 221.222

[2] Beginning Block Number 1. Method = Enter

The first footnote in Figure 8-8 shows that *–2LL* for the initial model is 221.222. For the model that contains age, histology, and stage of the disease, *–2LL* is 202.881. The difference between these two numbers is 18.340. This is the entry labeled *Change from Previous Step*. The degrees of freedom are the difference between the number of parameters in the two models. In this example, the degrees of freedom are 4, since you are comparing a model with no parameters to one with four parameters. The observed significance level is small—0.001—so you can reject the null hypothesis that all of the coefficients are 0.

You can specify the sequence in which one or more predictor variables are entered into the Cox regression model using the Block toggle in the Cox Regression dialog box. The change in *–2LL* between successive blocks of variables is labeled *Change from Previous Block*. In this example, the variables are entered in a single step, so the *Change from Previous Block* and *Change from Previous Step* values are equal.

Another test of the null hypothesis that all parameters are 0 can be based on the **score statistic**. This test is also sometimes called the **global chi-square** or the **overall chi-square**. Figure 8-8 shows that the value for the overall (score) chi-square is very similar to that of the change in the likelihood statistic. As expected, you reach the same conclusion.

Plotting the Estimated Functions

You can examine the estimated cumulative survival and cumulative hazard functions by plotting them at the mean values of the covariates or at values that you specify (patterns). You can also plot separate functions for each value of a variable.

Recall the Cox Regression dialog box and select:

Plots...
    Plot Type
    ☑ Hazard

    Covariate Values Plotted at: age(Mean), hist(Cat)(Mean)

    ▶ Separate Lines for: stage(Cat)

Cumulative Hazards At Average Values of Covariates for Each Stage

Figure 8-9 is a plot of the cumulative hazard function at the average values of the covariates. Two separate curves are shown—one for early-stage patients and one for late-stage patients. The shape of the curves is a result of the proportional hazards assumption. The ratio of the two hazards at any time points is equal. Both curves are evaluated at event times in either of the two groups. If one of the groups has a more restricted period of observation, this will result in extrapolation of the curve beyond the observed range of values—an undesirable feature.

Figure 8-9
Cumulative hazard by stage

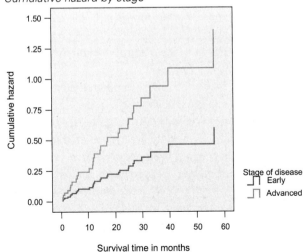

Cumulative Survival Plot for a Covariate Pattern

To plot the estimated cumulative survival for a 40-year-old with a histology of nodular sclerosis (coded 1) for the equation with age, stage, and histology, recall the Cox Regression dialog box and select:

Plots...
   Plot Type
   ☑ Survival

   Covariate Values Plotted at: age(40), hist(1), stage(Mean)

Save...
   Survival
   ☑ X*Beta

   Diagnostics
   ☑ Hazard function
   ☑ Partial residuals
   ☑ DfBeta(s)

This also saves diagnostic variables and plots for future use.

Figure 8-10
Estimated cumulative survival for 40-year-old with nodular sclerosis

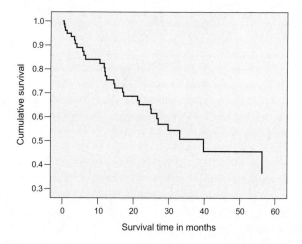

Checking the Proportional Hazards Assumption

The hazard function, like the survival function, is factored into two component pieces. The baseline hazard depends only on time, while the second component depends only on the values of the covariates and the coefficients. One of the assumptions of the Cox regression model is that for any two cases, the ratio of the estimated hazard across time is a constant. For example, if you have two patients with the same age and histology but different stages of the disease, the ratio of their estimated hazard rates across all time points is e^B, where B is the regression coefficient for stage. This is not an assumption to be made lightly. It is quite possible that the hazard functions of early- and late-stage patients are not related by a constant multiplier and that their ratio depends on time.

There are several ways to examine the proportional hazards assumption. If you have a categorical variable, you can estimate different baseline hazard functions for each category of the predictor variable. Then you can compare the baseline hazard functions. For both continuous and categorical variables, you can create an interaction term between time and the covariate of interest and determine whether it has a coefficient significantly different from 0. (For more information, see "Time-Dependent Covariates" on p. 165.)

Stratification

It is possible to modify the Cox regression model to incorporate separate baseline hazard functions for each value of a categorical variable or for grouped values of a continuous variable. (The groups for which separate models are estimated are called **strata**.) For example, if you have reason to believe that strata have different baseline hazard functions, you can fit the model

$$h_i(t) = h_{0,i}(t)e^{(BX)}$$

where $h_i(t)$ is the hazard function for stratum i. In this model, separate baseline hazard functions $h_{0,i}(t)$ are estimated for each stratum. The effects of the covariates are still assumed to be the same in all of the groups. Let's apply this model to the Hodgkin's disease data, using stage of disease as the strata variable.

Recall the Cox Regression dialog box and select:

▶ Covariates: age, hist

　Method: Enter

▶ Strata: stage

Categorical...
　▶ hist(Indicator)

Plots...
　Plot Type
　☑ Hazard
　☑ Log minus log

The SPSS Cox Regression procedure estimates two separate baseline hazard functions, one for each of the stages. Only one set of "pooled" coefficients for the predictor variables is estimated because you assume that the influence of the covariates is the same in all of the strata. There are no coefficients for the stratification variable. Its only effect is to identify the groups for which separate baseline hazards are estimated. You may want to determine whether the baseline hazard functions are proportional in the strata because if they are, the variable used to form the strata can be used as a covariate in the model, and you can estimate a common baseline hazard for both groups.

Log-Minus-Log Survival Plot

A useful plot for assessing whether the baseline hazard functions are proportional is called the **log-minus-log (LML) survival plot**. If the baseline hazard functions are proportional, the lines for the individual strata should be parallel—that is, the distance between them should be the same at all time points.

Figure 8-11 is the LML plot for the Hodgkin's disease data when *stage* (of disease) is the stratification variable and *age* and *hist* (histology) are covariates. You see that the two lines are more-or-less parallel and that it may be unnecessary to stratify the data based on *stage*. In the next section, you'll construct a formal test of the proportional hazards assumption using a time-dependent covariate (see p. 169).

Figure 8-11
Log-minus-log plot

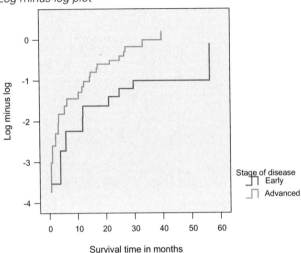

Identifying Influential Cases

You do not want the results of any statistical analysis to depend heavily on the values for an individual case, which is why it is always important to look for **influential cases** in your data. (A case is influential if there are substantial changes in the values of the coefficients when the model is computed with the case and without the case.) The Cox proportional hazard model uses only the rank ordering of the times, so outliers don't affect the estimation of the coefficients as much as they do in a parametric analysis.

A statistic that estimates the change in regression coefficients with and without a case is called **DfBeta**. For each case, you can compute a DfBeta for each variable in the model. Cases with outlying values for DfBeta should be examined. An easy way of examining the DfBetas is to plot for each variable in the model the DfBetas against the case-identification number.

Figure 8-12 is a plot of the estimated changes in the age coefficient when each of the cases in turn is excluded from the model that contains *age*, *hist*, and *stage*. (You can save the change in the coefficients by selecting DfBeta(s) in the Save New Variables dialog box. You can then create the scatterplot by choosing Scatter/Dot from the Graphs menu. Double-click the plot to open it in the Chart Editor, and click on points using the Data Label Mode tool.) Note that, as expected, most of the values fall in a horizontal band around 0.

One value is far removed from the rest—the value for case 24. If you exclude case 24 from the estimation of the model, the coefficient for *age* changes from 0.0301 to 0.0406, resulting in a DfBeta for *age* for case 24 of 0.0113. (The DfBetas calculated in the Cox Regression procedure are an approximation, so their values will not be identical to those obtained from actually removing the case from the analysis.) If you look at the data, you will see that case 24 is a 60-year-old who lived for a long time. (The average age for all of the patients is 38.73). This case decreases the value of the coefficient for *age* because, overall, as age increases, so does the probability of dying.

Whenever you find influential observations in your data, you should check the values for the case to make sure that they are not the result of coding or data-entry errors. If the recorded values are correct, when reporting the results of your analyses you should include discussion of the impact of the influential case on the results of the analyses. See Belsley et al. (1980) for further discussion.

Figure 8-12

DfBeta against case-identification number

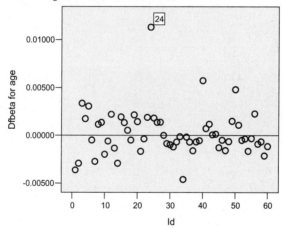

Examining Residuals

Several types of residuals have been found to be useful in detecting violations of the proportional hazards assumption. They are not as straightforward as the difference between observed and predicted values found in regression analysis, since there are no observed and predicted values that fit the usual definitions.

Partial (Schoenfeld) Residuals

The **partial residual**, suggested by Schoenfeld (1982), is computed for each covariate in the model. When you select Partial residuals in the Save New Variables dialog box, a partial residual is saved for every variable in the model. For each uncensored case, the partial residual for a covariate is the difference between the observed value of the covariate and its conditional expectation based on the cases still under observation when the case fails.

You can plot partial residuals against time to test the proportional hazards assumption. Crowley and Storer (1983) discuss some of the problems associated with plotting the partial residuals against the values of the covariates. Figure 8-13 is a plot of the partial residuals for age against survival time with a loess smooth. (These are the values saved from the model for age, histology, and stage. The partial residual for age is named *PR1_1*). Notice that the values appear to be more-or-less randomly distributed in a band around 0.

Figure 8-13
Partial residuals for age against survival time

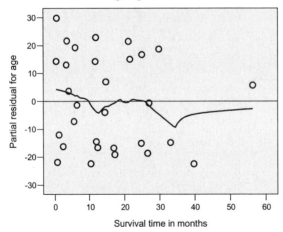

Martingale Residuals

Another type of residual used in proportional hazards models is the **martingale residual** (Therneau et al., 1990). Although it is not directly saved in the SPSS Cox Regression procedure, it can be computed easily by saving the cumulative hazard. If the event is coded as 1 and censored cases are coded as 0, the martingale residual is the difference between the observed event code and the predicted cumulative hazard (sometimes called the Cox-Snell residual).

In the Compute Variable dialog box (choose Compute from the Transform menu), enter:

martingale=event-haz_1

The sum of the martingale residuals is 0. The largest possible value is 1. They are not symmetric about 0. The martingale residuals can be plotted against the values of the independent variables or the linear predictor score (X'Beta), which is the sum of the products of the coefficients and the mean-corrected variables for a case.

Figure 8-14 is a plot of the martingale residuals against the values of age. Most of the points fall in a horizontal band around 0, which is the result you want to see. However, there is one outlying case with a martingale residual close to –3. As you would expect, this is case 24 again. The plot of the martingale residuals against the linear predictor scores (see Figure 8-15) again shows case 24 as an outlier. You don't want to see a relationship between the martingale residuals and the linear predictor scores.

Figure 8-14
Martingale residuals against age

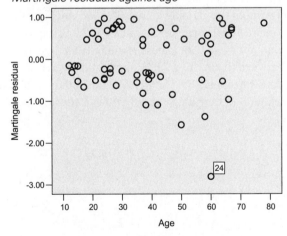

Figure 8-15
Martingale residuals against linear predictor scores

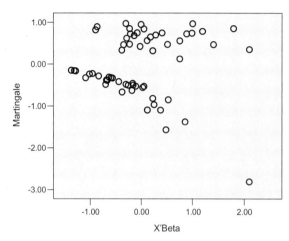

Examining the Functional Form of a Covariate

For continuous variables, martingale residuals are useful for checking whether the relationship between the log of the hazard and the values of the variable are linear. To do this, you fit the model without the variable in question and calculate the martingale residuals. Then you plot the residuals against the values of the variable.

Figure 8-16 shows the plot for *age* with a loess smooth. If the linearity assumption is met, the relationship between the martingale residuals and the variable should be approximately linear. This seems plausible for the example, although if you have reason to suspect that the hump in the plot is "real," you can consider an alternative form for *age*.

Figure 8-16
Martingale residuals against age

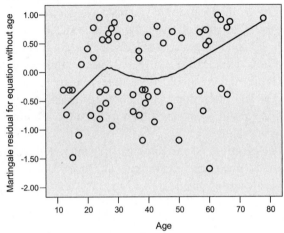

Selecting Predictor Variables

When you have many variables and do not know which, if any, are related to survival, you may want to build a model that excludes covariates that do not appear to be good predictors. The usual methods for variable selection can be used to build models in Cox regression. All of the problems associated with variable-selection algorithms remain. None of the algorithms results in a "best" model in any true sense. It is a good idea to examine several possible models and choose among them on the basis of interpretability, parsimony, and ease of variable acquisition. Usually, the model will fit the sample from which it is estimated better than it will fit the population from which the sample is selected. Another sample from the same population will often result in a different model. Correlated covariates may result in variables that are individually highly related to outcome being excluded from the model. The significance levels for variable entry and removal cannot be interpreted in the usual manner.

Variable Selection Methods

The Cox Regression procedure provides several methods for model selection. All variables that represent the same categorical variable are entered or removed from the model in the same step.

You can enter all variables into a model in a single step using the **forced-entry method**, or you can specify the sequence in which groups of variables are entered by using the Block toggle. This method is useful if you want to enter a core set of variables into the model and then use a variable selection algorithm to suggest other predictors.

You can also use forward and backward variable selection for model building. In **forward stepwise selection**, variables are considered one at a time for entry into the model. After a variable is added to a model, all variables already in the model are examined for removal. The algorithm stops when no more variables meet entry or removal criteria or when the resulting model is identical to a previous one (which means that the algorithm is cycling).

In **backward stepwise selection**, all variables are first entered into the model in a single step. Then the variables are examined for removal. Once no more variables meet removal criteria, variables are again considered for entry. The algorithm stops when no more variables meet entry or removal criteria.

Entry and Removal Criteria

The score statistic is always used for entering variables into a model. One of three statistics can be used for variable removal: the likelihood-ratio statistic based on the maximum partial likelihood estimates, the likelihood-ratio statistic based on conditional parameter estimates (Lawless and Singhal, 1978), and the Wald statistic. The likelihood-ratio statistic based on the maximum partial likelihood estimates is probably the best criterion for variable removal; however, it is computationally intensive because it requires repeated model reestimations. The change in log-likelihood is computed by deleting each variable in turn from the model and calculating the change in the log-likelihood. The statistic based on conditional estimates does not require reestimation of the model and, in many situations, tends to give similar results to the statistic based on the maximum partial likelihood estimates.

An Example of Forward Selection

To see what the output looks like when a model is built using forward selection, consider the Hodgkin's disease data again. The predictor variables that you will consider for entry into the model are *age*, *sex*, *stage*, and *hist* and the two-way interactions among *sex*, *stage*, and *hist* (*sex*, *stage*, and *hist* are declared as categorical covariates with deviation parameter coding). The likelihood-ratio test based on the partial likelihood estimates is used for variable removal.

When interaction terms are included in the model, they can cause large changes in the statistics for the main effects, since the main-effects estimates have different meanings. If you have interaction terms, you want to include the corresponding main effects as well so that you can interpret the model properly. To include interaction terms, select (in the source list) all of the variables involved in the interaction, and then click ≥a*b>.

Recall the Cox Regression dialog box and select:

▶ Covariates: age, sex, stage, hist, sex*stage, hist*stage, hist*sex

Method: Forward:LR

▶ Categorical: sex(Deviation), stage(Deviation), hist(Deviation)

Figure 8-17 shows the coding used for the categorical variables. You selected deviation coding, so all terms are easily interpretable.

Figure 8-17
Categorical variable coding

Categorical Variable Codings[2,3,4]

		Frequency	(1)	(2)
sex[1]	0=female	18	1	
	1=male	42	-1	
stage[1]	0=Early	26	1	
	1=Advanced	34	-1	
hist[1]	1=nodular sclerosis	32	1	0
	2=mixed cellular	21	0	1
	3=lymphocyte depletion	7	-1	-1

[1.] Deviation Parameter Coding

[2.] Category variable: sex (sex)

[3.] Category variable: stage (stage of disease)

[4.] Category variable: hist (histology)

Selecting the First Variable

At each step of model building, variables that are not in the model are considered for inclusion based on the observed significance level for the score statistic. Figure 8-18 shows the score statistic and its observed significance level for each of the candidate variables. For categorical variables and interactions with more than one degree of freedom, an overall score statistic for the variable as well as the score statistics for the individual components are shown. The overall score statistic determines whether the variable is entered. The smallest observed significance level is for the histology variable. Since it is less than 0.05, the default for variable entry, histology is the first variable to enter the model.

Figure 8-18
Variables not in the equation at step 0

Variables not in the Equation[1]

	Score	df	Sig.
age	7.104	1	.008
sex	.030	1	.862
stage	4.143	1	.042
hist	12.797	2	.002
hist(1)	6.279	1	.012
hist(2)	2.653	1	.103
sex*stage	.169	1	.681
hist*stage	.829	2	.661
hist(1)*stage	.582	1	.446
hist(2)*stage	.337	1	.562
hist*sex	3.824	2	.148
hist(1)*sex	1.005	1	.316
hist(2)*sex	2.721	1	.099

[1] Residual Chi Square = 52.908 with 10 df Sig. = .000

The residual chi-square, displayed below the table of score statistics, is a test of the null hypothesis that the coefficients for all variables *not* in the model are 0. If the observed significance level for the residual chi-square is small, you can reject the hypothesis that all of the coefficients are 0. In this example, the observed significance level is less than 0.0005, so it appears that some of the variables not in the model are related to survival.

Coefficients for Variables in the Model At Step 1

The estimated deviation coefficients for the histology variable are shown in Figure 8-19. The first two histology types (nodular sclerosis and mixed cellular) are associated with decreased death rates because the coefficients are negative. This means that the third category, which is not shown, is associated with increased death rates. (Its coefficient is 1.006, since with deviation contrasts the sum of all of the coefficients is 0.)

Figure 8-19
Statistics for model containing histology

		B	SE	Wald	df	Sig.	Exp(B)
Step 1	hist			10.652	2	.005	
	hist(1)	-.591	.251	5.540	1	.019	.554
	hist(2)	-.415	.272	2.320	1	.128	.660

Evaluating Variables for Entry At Step 2

To decide which variable enters the model at step 2, look at the statistics shown in Figure 8-20. The smallest observed significance level is for the variable *age*, so it is entered into the model containing histology.

Figure 8-20
Variables not in the model after step 1

Variables not in the Equation[1]

		Score	df	Sig.
Step 1	age	5.762	1	.016
	sex	.588	1	.443
	stage	4.076	1	.043
	sex*stage	.397	1	.529
	hist*stage	.660	2	.719
	hist(1)*stage	.525	1	.469
	hist(2)*stage	.349	1	.554
	hist*sex	6.192	2	.045
	hist(1)*sex	1.743	1	.187
	hist(2)*sex	5.783	1	.016

[1] Residual Chi Square = 25.348 with 8 df Sig. = .001

Evaluating Variables for Removal At Step 2

Figure 8-21 contains removal statistics for the two variables in the model at step 2. You can base the decision to remove a variable on three different statistics: changes in likelihood based on the maximum partial likelihood estimates, changes in likelihood based on the conditional likelihood estimates, and the Wald statistic. In Figure 8-21, the changes in likelihood are based on the conditional parameter estimates. Since both of the variables have observed significance levels of less than 0.1 (the default for variable removal), neither of them can be removed and the algorithm continues screening the variables for entry.

162

Chapter 8

Figure 8-21
Removal statistics for age and histology

Term Removed		Loss Chi-square	df	Sig.
Step 2	age	5.664	1	.017
	hist	6.861	2	.032

Evaluating Variables At Step 3

The next variable to be entered is stage of the disease, since it has the smallest observed significance level for the score statistic of all variables that are not in the model, as shown in Figure 8-22.

Figure 8-22
Variables not in the model after step 2

		Score	df	Sig.
Step 2	sex	.400	1	.527
	stage	4.485	1	.034
	sex*stage	.323	1	.570
	hist*stage	3.954	2	.139
	hist(1)*stage	2.586	1	.108
	hist(2)*stage	2.269	1	.132
	hist*sex	5.100	2	.078
	hist(1)*sex	2.981	1	.084
	hist(2)*sex	3.769	1	.052

Statistics for the variables in the equation are shown in Figure 8-23. The coefficient for *stage* is negative, indicating that the early stage of the disease has a better prognosis than the later stage. Since *stage* is a categorical variable with only two values (early and late), it has one degree of freedom and only one coefficient is printed. The coefficient corresponds to deviation parameter estimates, so it tells you the effect of the early stage compared to the average effect of both stages.

Figure 8-23
Statistics for the model containing age, hist, and stage

		B	SE	Wald	df	Sig.	Exp(B)
Step 3	age	.030	.012	5.831	1	.016	1.031
	stage	-.424	.205	4.262	1	.039	.655
	hist			6.309	2	.043	
	hist(1)	-.348	.273	1.629	1	.202	.706
	hist(2)	-.483	.272	3.153	1	.076	.617

From the observed significance level for the loss chi-square shown in Figure 8-24, you see that none of the variables is eligible for removal because the observed significance level is less than 0.1 (the default) for each of them.

Figure 8-24
Removal statistics at step 3

	Term Removed	Loss Chi-square	df	Sig.
Step 3	age	6.091	1	.014
	stage	4.636	1	.031
	hist	5.246	2	.073

From the statistics for variables not in the equation shown in Figure 8-25, you see that none of the variables meets the entry criteria. Since components of categorical variables are all entered or removed together, only the significance for the overall variable is considered for entry and removal. (If you want the component variables of a categorical variable to be entered and removed individually, you must use the transformation language outside of the Cox Regression procedure to create the component variables; then you can use these variables as you would any others in the model.) Forward model selection stops at this point because no more variables are eligible for entry or removal.

Figure 8-25
Statistics for variables not in the equation

Variables not in the Equation[3]

		Score	df	Sig.
Step 3	sex	.289	1	.591
	sex*stage	.019	1	.892
	hist*stage	2.005	2	.367
	hist(1)*stage	.695	1	.405
	hist(2)*stage	1.394	1	.238
	hist*sex	5.067	2	.079
	hist(1)*sex	2.533	1	.112
	hist(2)*sex	4.160	1	.041

[3]. Residual Chi Square = 12.128 with 6 df Sig. = .059

Omnibus Test of the Model At Each Step

Each time a variable is entered into the model, you can test whether the coefficients for all variables in the model are 0 and whether the coefficients for the variable added at that step are 0. Based on changes in $-2LL$, you can test the hypothesis that the regression coefficients for both age and histology are 0 and that the coefficient for *age* is 0 when the variable is added to a model that already contains histology.

Figure 8-26
Omnibus test of model coefficients at each step

Omnibus Tests of Model Coefficients [4,5]

		Step		
		1 [1]	2 [2]	3 [3]
-2 Log Likelihood		213.181	207.517	202.881
Overall (score)	Chi-square	12.797	19.130	24.980
	df	2	3	4
	Sig.	.002	.000	.000
Change From Previous Step	Chi-square	8.041	5.664	4.636
	df	2	1	1
	Sig.	.018	.017	.031
Change From Previous Block	Chi-square	8.041	13.705	18.340
	df	2	3	4
	Sig.	.018	.003	.001

[1] Variable(s) Entered at Step Number 1: hist

[2] Variable(s) Entered at Step Number 2: age

[3] Variable(s) Entered at Step Number 3: stage

[4] Beginning Block Number 0, initial Log Likelihood function: -2 Log likelihood: 221.222

[5] Beginning Block Number 1. Method = Forward Stepwise (Likelihood Ratio)

In Figure 8-26, the *–2LL* for the model without any predictor variables is 221.22. For the model with histology alone, the *–2LL* is 213.18. The change, which has a chi-square distribution with two degrees of freedom, is 8.04, which is the *Change from Previous Block* at step 1. It's also the change from the previous step, since it's the first step. You can reject the null hypothesis that the coefficients for the histology variables are 0.

For the model with both age and histology, *–2LL* is 207.52. The change in *–2LL* when age is added to a model that contains histology is $213.18 - 207.52$, or 5.66, which is the chi-square value labeled *Change from Previous Step* in Figure 8-26. Since the observed significance level is small, you can reject the null hypothesis that the coefficient for age is 0.

If you build a model by sequentially specifying different steps, such as by entering *age* into a model containing no variables and then performing forward stepwise selection using the remaining variables as candidates for entry and removal, each of the methods corresponds to a block. The current model is always compared to the final model from the previous block. If you do not have multiple blocks, the previous block is the model with no variables.

Time-Dependent Covariates

When people are followed for an extended length of time, the values of many variables may change. For example, blood pressure, weight, hemoglobin, and dosages of drugs administered may fluctuate over the course of a study. Even variables like marital status or type of treatment administered need not be constant.

Any predictor variable whose values change with time is known as a **time-dependent covariate**. Such covariates are easily incorporated into Cox regression models. However, the actual computations take much longer because the values of the covariates must be generated at each time point.

Time-dependent covariates may be categorical (for example, on the drug, off the drug) or continuous (for example, laboratory values). Since the time-dependent covariate value must be defined for all observed time points in the study, continuous covariate values may be assumed to be constant between two measurement times. That is, if you measure blood pressure at time 1 and then again at time 2, you may want to assume that the blood pressure values between times 1 and 2 are equal to those at time 1, or the average of the two values. You'll have to make some assumptions, since you have to define the values of the covariates at all time points.

To see how you can analyze data when hazards are not proportional, consider the following example presented by Stablein et al. (1981). They report survival times for two groups of patients with locally advanced, nonresectable gastric carcinoma. One group of patients was treated using a combination of chemotherapy and radiation, while the other group was treated with radiation alone. The question of interest is whether the two treatments are equally effective in prolonging survival.

Examining the Data

Before you start to build a model, you should always look at preliminary plots that describe the data. You can do this in either the Kaplan-Meier or Cox Regression procedure. If you don't have any covariates, the baseline cumulative survival at the mean value of the covariates displayed in the Cox Regression procedure is the same as the cumulative survival in the Kaplan-Meier procedure.

To obtain cumulative survival and log-minus-log plots for the two treatments, open the data file *gastricancer.sav*, and from the menus choose:

Analyze
 Survival
  Cox Regression...

▶ Time: time

▶ Status: event

Define Event...
  Single value: 1

▶ Strata: treat

Plots...
  Plot Type
  ☑ Survival
  ☑ Log minus log

Figure 8-27 contain a plot of the cumulative survival function for the two treatment groups. Notice that the two survival curves cross. This is an indication that the proportional hazards assumption may not be appropriate for these data.

Figure 8-27
Cumulative survival functions for treatment groups

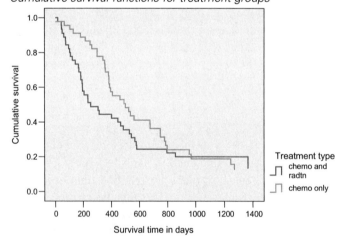

Figure 8-28
Log-minus-log plot of hazard functions for treatment groups

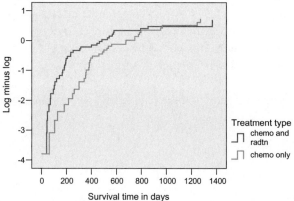

The log-minus-log plot shown in Figure 8-28 also suggests that the hazards in the two groups are not proportional. The points for the two treatment groups do not appear to fall on parallel lines. Since the main question of interest in this analysis is whether survival differs for the two treatments, you need a model that incorporates nonproportional hazards over time and allows you to estimate the treatment effect. A simplification of the model suggested by Stablein et al. (1981) is

$$h(t) = h_0(t)e^{B_1 * treat + B_2 * treat * t\_cov\_}$$

This model differs from those previously considered because time is included as a predictor variable. The above model contains two predictor variables: treatment (*treat*), (coded as 0 for chemotherapy and radiation and 1 for radiation alone) and treatment multiplied by time (*treat*t_cov_*). This type of model does not force the ratio of the hazard rates for the two groups to be constant over time. Instead, the ratio can vary over time.

Specifying a Time-Dependent Covariate

To set up a Cox time-dependent analysis, from the menus choose:

Analyze
  Survival
    Cox w/ Time-Dep Cov...

A new variable, named *T_*, has been inserted at the top of the variables list (on the left in the dialog box). This variable is generated by the time-dependent procedure. It assumes the same values as the variable in the database that is designated as the survival *time* indicator. $T\_$ can be transformed in many ways. For example, if $T\_$ is time in months, $T\_ / 12$ would measure years. After you enter the information in the dialog box, the transformed version is stored in a variable named $T\_COV\_$. The strategy is to use $T\_COV\_$ as a covariate in interaction with other variables to generate a regression equation in which variables are free to change as a function of time.

To create a time-dependent covariate for this example (time in months instead of days), specify:

Expression for T_COV_: T_/30

and then select:

Model...

▶ Time: time

▶ Status: event

Define Event...
  Single value: 1

▶ Covariates: treat, T_COV_*treat (select *T_COV_* and *treat* and click ≥a*b>)

The term $T\_COV*treat$ is the treatment-by-time interaction.

Figure 8-29 contains the output from fitting the model with the treatment-by-time interaction. From the coefficient for treatment and treatment-by-time and their observed significance levels, you see that there is a treatment effect and that it is not constant over time. (If it were constant over time, you would not reject the null hypothesis that the treatment-by-time coefficient is 0.) The chemotherapy treatment is better for early time points, but its superiority decreases over time, since the coefficient for treatment-by-time is positive.

Figure 8-29

Coefficients for the time-dependent Cox model

	B	SE	Wald	df	Sig.	Exp(B)
treat	-1.209	.431	7.871	1	.005	.299
T_COV_*treat	.071	.028	6.597	1	.010	1.074

Calculating Segmented Time-Dependent Covariates

If you have a covariate whose values change with time, you must use the time variable to define its value at each time point. For example, if you measure tumor size at the start of the study and then at two months and three months, for each case you have three tumor measurements, such as *size0*, *size2*, and *size3*. Tumor size is then a time-dependent covariate. You have to match its values with the times in the study. For example, you want to use *size0* for times less than two months, *size2* for times between two months and three months, and *size3* for three or more months. You can define the time-dependent covariate as

$$T\_COV\_ = (T\_ < 2)*size0 + (T\_ \geq 2 \& T\_ < 3)*size2 + (T\_ \geq 3)*size3$$

The expressions in parentheses will have values of 0 or 1, depending on the value of time. If time is less than two months, the value used for $T\_COV\_$ is *size0*, since only the value of the first expression in parentheses is 1. If time is greater than or equal to two months but less than three months, $T\_COV\_$ is *size2*. If time is greater than or equal to three months, $T\_COV\_$ is *size3*.

The dialog box interface restricts the complexity of the time-dependent covariates that can be specified, since you can calculate only one variable. However, using the command syntax, you can specify considerably more complicated structures.

Testing the Proportional Hazard Assumption with a Time-Dependent Covariate

Whenever you want to test that hazards are proportional for different strata, you can incorporate a time-by-stratification-variable interaction. If you can reject the null hypothesis that the coefficient for this term is 0, there is sufficient reason to believe that the hazards are not proportional.

In the previous example, when you modeled the relationship between survival, age, histology and stage, you examined plots to check the proportional hazards assumption for the two stages. You can also test the assumption by including an interaction term between time and stage.

Reopen the data file *hodgkins.sav*, and from the menus choose:

Analyze
  Survival
    Cox w/ Time-Dep Cov...

▶ Expression for T_COV_: T_

Model...
    ▶ Time: survivaltime
    ▶ Status: dead

Define Event...
    Single value: 1

▶ Covariates: age, hist, stage, T_COV_*stage (select *T_COV_* and *stage* and click ≥a*b>)

Categorical...
    ▶ hist(Indicator)

The coefficients are shown in Figure 8-30. You see that the stage-by-time interaction term is not significant, lending support to the assumption that the hazards are proportional between the two stages. You can also test interactions of continuous variables with time in the same manner. For example, you can create the interaction of *T_COV* and *age*.

Figure 8-30
Testing proportional hazards with time-dependent covariate

	B	SE	Wald	df	Sig.	Exp(B)
age	.030	.012	5.814	1	.016	1.031
hist			6.129	2	.047	
hist(1)	-1.167	.546	4.568	1	.033	.311
hist(2)	-1.297	.545	5.660	1	.017	.273
stage	1.099	.658	2.787	1	.095	3.001
T_COV_*stage	-.015	.030	.254	1	.614	.985

Fitting a Conditional Logistic Regression Model

The Cox Regression procedure can be used to fit a conditional logistic regression model, which is used to analyze the occurrence of an event when each case is paired with one or more matched controls. The Multinomial Logistic Regression procedure can also be used to analyze the situation when each case has exactly one control. However, the data file must be transformed prior to analysis. The same data used in Chapter 3 are reanalyzed in this section, leading to the same results but with less manipulation of the data (especially of categorical variables).

Data File Structure

To fit the conditional logistic regression model using the Cox Regression procedure, there must be a separate data record for each case and for each control. The data file must contain all of the risk factors of interest as well as the following variables:

- A unique number that is the same for a case and all of its controls. For the example, this variable is named *set*.

- A variable with two values that distinguishes cases and controls. All cases have the same value. All controls have the same value. The number must be positive, and it must be larger for the controls. For the example, this variable is named *outcome*, cases have a value of 1, and controls have a value of 2.

- A copy of the previously described variable (*outcome*). For the example, the variable is named *outcome2*.

The same pair of data records shown in Table 3-1 in Chapter 3 are shown in Figure 8-31.

Figure 8-31
Data file for conditional logistic regression

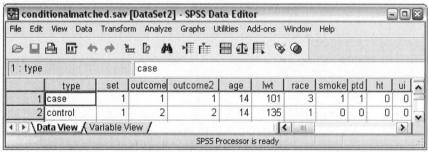

Specifying the Analysis

To obtain the analysis, from the menus choose:

Analyze
  Survival
    Cox Regression...

▶ Time: outcome

▶ Status: outcome2

Define Event...
  Single value: 1

▶ Covariates: smoke, ui, ptd, race, lwt

Categorical...
  Categorical Covariates
   ▶ race(Indicator)

  Reference Category: First

▶ Strata: set

Race is declared as a categorical variable with indicator coding. The first category is chosen as the reference category to make the results comparable to those shown in Figure 3-15 in Chapter 3.

Parameter Estimates

The parameter estimates from the Cox Regression procedure are shown in Figure 8-32. They are identical to those shown in Figure 3-15 in Chapter 3. Although this example was for a single control for each case, multiple controls for a single case are analyzed in exactly the same way. However, you should not analyze data with multiple cases and multiple controls in a set with the Cox Regression procedure, given the method implemented for handling tied survival times within a stratum.

Figure 8-32
Matched-pair parameter estimates

Variables in the Equation

	B	SE	Wald	df	Sig.	Exp(B)
smoke	1.348	.568	5.631	1	.018	3.849
ui	1.032	.664	2.418	1	.120	2.808
ptd	1.563	.706	4.897	1	.027	4.774
race			1.849	2	.397	
race(1)	-.469	.604	.603	1	.438	.626
race(2)	.392	.673	.339	1	.560	1.480
lwt	-.009	.009	1.051	1	.305	.991

Variance Components

What components contribute to variation in the absorption of calcium by turnip leaves? Can you estimate the size of the individual components? Is variance in testing of ovens due to their nonhomogeneity? These questions can be investigated by using the Variance Components procedure. In this chapter, you can find illustrations of four different methods, as well as technical background information.

Factors, Effects, and Models

A key concept in the variance components model is the idea of fixed and random effects. Factors make up the effects, and then the effects are combined into a model.

Types of Factors

Since effects in the model are composed of factors, first consider the types of factors. When the levels of a factor are all of the possible values in the entire population or all of the levels of interest to the researchers, then it is a **fixed factor**. For example, imagine that researchers for a computer magazine conduct a study to compare the lifetimes of four types of laptop computer batteries: nickel cadmium (NiCad), nickel metal hydride (NiMH), lead acid, and lithium ion (LiIon). The variable labeled *Type of Battery* is treated as a fixed factor either when these four types are the only ones on the market or when the researchers are interested in making inferences about only these four types.

A factor is said to be **random** when the levels of the factor represent a random sample of all possible values from a population and inferences will be made on the entire population. Suppose, in the above example, that the study of battery types is conducted using batteries from four different manufacturers. It is not unreasonable to think of those manufacturers as a randomly chosen sample from a population of battery manufacturers. The lifetime of batteries from any one of these four manufacturers may be of no particular interest in and of itself to the researchers. The four manufacturers are chosen with the objective of treating them as representative of the population of all laptop computer battery manufacturers, and inferences can and will be made about the whole population. Thus, the variable labeled *Manufacturer* is treated as a random factor.

Types of Effects

Effects are made up of factors, and in the Variance Components procedure, they also inherit the random nature of the composing factors. An effect is treated as a **fixed effect** when all of its composing factors are fixed factors; otherwise, the effect is treated as a **random effect**.

In the laptop computer battery example, the main effect *Type of Battery* is a fixed effect, and the nested effect *Manufacturer within Type of Battery* is a random effect.

Types of Models

If all effects in a model are random effects, the model is called a **random-effects model**. A model with only fixed effects and the residual term is called a **fixed-effects model**. It is common to find models where some effects are random and some are fixed. In this case, the model is a **mixed model**. Since the intercept is generally treated as a fixed effect, a model including the intercept and one or more random factors is by default a mixed model. Using the above laptop computer battery example, an example of a mixed model is

Intercept + *Type of Battery* + *Manufacturer within Type of Battery* + Residual

while an example of a fixed-effects model is

Intercept + *Type of Battery* + Residual

Model for One-Way Classification

For a first example of variance components, classification by one factor provides a simple model. This section deals with the one-way classification model generally, using one data set to demonstrate two methods of variance component estimation. Consider the turnip leaf data that appeared in Table 13.3.1 of Snedecor and Cochran (1980). As mentioned in the authors' book, the data presented here are a small subset of the original data collected from a larger experiment on the precision of estimation of calcium concentration in percentage dry weight of turnip greens. Four determinations were made on each of the four leaves from a single plant. The data are shown in Figure 9-1 and are in the file *Turnip leaves.sav*.

Figure 9-1
Calcium concentration in turnip greens (% dry weight)

Leaf	% Calcium Determination			
	1	2	3	4
1	3.28	3.09	3.03	3.03
2	3.52	3.48	3.38	3.38
3	2.88	2.80	2.81	2.76
4	3.34	3.38	3.23	3.26

In the data file *Turnip leaves.sav*, each measurement is one case. The variables are leaf (*leaf*), percentage of calcium (*pctca*), and determination (*determ*).

Consider how the mean calcium concentration varies with *leaf*. You can use the Case Summaries procedure to compute the observed means and the observed standard deviations for each level of *leaf*. The results are shown in Figure 9-2.

Figure 9-2
Case summaries

Percentage of Calcium

Leaf	Mean	Std. Deviation
1	3.1075	.1184
2	3.4400	.0712
3	2.8125	.0499
4	3.3025	.0695
Total	3.1656	.2540

It is apparent that the means vary among these four leaves; thus, *leaf* is a plausible effect. Now the question is: Are you really interested in (1) the different percentages of calcium in the four specific leaves chosen or (2) the variation in all leaves from which the four leaves may have been drawn? Since the leaf *i* $(i = 1, 2, 3, 4)$ is just one from among randomly picked leaves, the answer should be 2, and the *leaf* effect must be treated as a random effect.

Let y_{ij} be the *j*th determination on leaf *i*. The mathematical model that relates the expected percentage on leaf *i* to the grand mean and the *leaf* effect is

$$y_{ij} = \mu + \alpha_i + \varepsilon_{ij} \qquad \text{for } j = 1, 2, 3, 4$$

where μ is the grand mean, α_i are the main effects $(i = 1, 2, 3, 4)$, and ε_{ij} are residual errors. The errors are independent, and each follows a distribution with a mean of 0 and variance of *Var(Error)*. For readers who are familiar with the general linear model, this equation looks exactly like the one for a one-way ANOVA model. However, some underlying assumptions are different: The main effects (α_i) are random variables whose values depend on which four leaves are picked. Thus, they follow certain probability distributions. It is usually assumed that the α_i are independent, and each has a mean of 0 and the same variance, *Var(leaf)*. Moreover, the main effects and the residuals are also assumed to be independent. It follows from these assumptions that

$$\text{Var}(y_{ij}) = \text{Var(leaf)} + \text{Var(Error)}$$

The total variance, then, of y_{ij} is the sum of the two components, *Var(leaf)* and *Var(Error)*—hence the name **variance components**. If the variable *leaf* has no effects on the percentage of calcium, then the component *Var(leaf)* equals 0. You can use the GLM Univariate procedure to test this null hypothesis (discussed in the *SPSS 14.0 Statistical Procedures Companion* [Norušis, 2005]). However, GLM cannot estimate the variance *Var(leaf)*, while the Variance Components procedure does estimate the variance.

Estimation Methods

Four estimation methods are available in the Variance Components procedure: ANOVA, minimum norm quadratic unbiased estimator (MINQUE), maximum likelihood (ML), and restricted maximum likelihood (REML). For illustration purposes, both the ANOVA method using Type I sums of squares and the maximum-likelihood method are used in this example.

ANOVA Method

The ANOVA method does not require any knowledge of the distributions of the random effect *leaf* and the residual error other than their means and the variances. This method estimates the variance components by equating the expected mean squares of the random effects and the residual to their observed mean squares. The expected and the observed mean squares vary depending on the type of sum of squares used. The Variance Components procedure offers two types of sums of squares: Type I and Type III. For a one-way classification model (one factor), both types produce the same sums of squares. For the sake of simplicity, only Type I results are shown here.

To produce the output, from the menus choose:

Analyze
  General Linear Model
    Variance Components...

▸ Dependent variable: pctca
▸ Random factor: leaf

Options...
  Method
   ⊙ ANOVA
  Sum of Squares
   ⊙ Type I
  Display
   ☑ Sums of squares
   ☑ Expected mean squares

The expected mean squares table is shown in Figure 9-3, and the ANOVA table is shown in Figure 9-4.

Figure 9-3
Expected mean squares

Source	Variance Component		Quadratic Term
	Var(leaf)	Var(Error)	
Intercept	4.000	1.000	Intercept
leaf	4.000	1.000	
Error	.000	1.000	

Dependent Variable: pctca
Expected Mean Squares are based on Type I Sums of Squares.
For each source, the expected mean square equals the sum of
the coefficients in the cells times the variance components, plus
a quadratic term involving effects in the Quadratic Term cell.

Figure 9-4
ANOVA table

Source	Type I Sum of Squares	df	Mean Square
Corrected Model	.888	3	.296
Intercept	160.339	1	160.339
leaf	.888	3	.296
Error	.079	12	.007
Total	161.307	16	
Corrected Total	.968	15	

Dependent Variable: pctca

Since the Variance Components procedure does not perform hypothesis testing, only the sums of squares, degrees of freedom, and mean squares are reported in the ANOVA table.

Reading from the expected mean squares table (Figure 9-3), the expected mean squares of *leaf* and *Error* are

$$\text{EMS(leaf)} = 4 \times \text{Var(leaf)} + \text{Var(Error)}$$
$$\text{EMS(Error)} = \text{Var(Error)}$$

Equating these expected mean squares with the corresponding observed mean squares in the ANOVA table produces the following two equations:

$$0.296 = 4 \times \text{Var(leaf)} + \text{Var(Error)}$$
$$0.006602 = \text{Var(Error)}$$

Solving for the variances yields

$$\text{Var(Error)} = 0.006602$$
$$\text{Var(leaf)} = \left(\frac{0.296 - 0.006602}{4}\right) = 0.0723$$

These solutions are the ANOVA Type I variance estimates. SPSS calculates these variances and displays full precision values of the estimates in the variance estimates table (Figure 9-5).

Figure 9-5

Variance estimates

Component	Estimate
Var(leaf)	.07238
Var(Error)	.00660

Dependent Variable: pctca
Method: ANOVA (Type I Sum of Squares)

The ANOVA Type I variance estimates are *Var(leaf)* = 0.07238 and *Var(Error)* = 0.006602, which agree with the estimates given in Table 13.3.1 in Snedecor and Cochran (1980).

Maximum-Likelihood Method

The maximum-likelihood method requires distributional assumptions other than those on the means and the variances of the main effects *leaf* and the residual error, which were the assumptions required for the ANOVA method. Maximum likelihood assumes that the data are normally distributed.

Under the normality assumption, the likelihood function is maximized with respect to a set of parameters. In practice, the natural logarithm of the likelihood function (called the log-likelihood function) is maximized. The parameters include all of the fixed effects and all of the variance components. In the turnip-leaf example, there are three parameters: the intercept term, the variance *Var(leaf)*, and the residual variance *Var(Error)*. The parameter values at which the log-likelihood function attains its maximum value are called the maximum-likelihood estimates. The Variance Components procedure uses an iterative algorithm to produce maximum-likelihood estimates. This algorithm combines the rapid convergence property of the Newton-Raphson method with the robust ability of the Fisher scoring method to correct inaccurate initial parameter estimates.

Although the complete set of parameters includes the intercept term, its estimate is not displayed. The variance estimates table in Figure 9-6 displays the maximum-likelihood estimates for the two variance components.

To produce this output, recall the dialog box and choose:

Options...
  Method
  ⊙ Maximum likelihood
  Display
  ☑ Iteration history

Figure 9-6
Variance estimates

Component	Estimate
Var(leaf)	.05387
Var(Error)	.00660

Dependent Variable: pctca
Method: Maximum Likelihood Estimation

The maximum-likelihood variance estimates are $Var(leaf) = 0.05387$ and $Var(Error) = 0.006602$. Comparing with the ANOVA Type I estimates, we find that the two methods give different estimates for $Var(leaf)$. This difference is expected, primarily because the maximum-likelihood method assumes normality in addition to the assumptions required by the ANOVA method.

Another reason that the maximum-likelihood method is sometimes preferred to the ANOVA method is that it readily gives an asymptotic variance-covariance matrix of the variance estimates as a by-product of the iteration procedure. In contrast, the sampling variance-covariance matrix of the ANOVA method estimates is often very difficult to derive.

The Variance Components procedure always displays the asymptotic covariance matrix table along with the variance estimates table. Using the asymptotic normality property of the maximum-likelihood estimators, the asymptotic variance-covariance matrix can be used in establishing confidence intervals and testing hypotheses about the variance components.

Figure 9-7
Asymptotic covariance matrix

	Var(leaf)	Var(Error)
Var(leaf)	.00154	-1.82E-006
Var(Error)	-1.82E-006	7.26E-006

Dependent Variable: pctca
Method: Maximum Likelihood Estimation

For example, variance for the estimated $Var(leaf)$, as read from Figure 9-7, is 0.00154. Taking the positive square root of this number gives the standard error for $Var(leaf)$, which is 0.03927. Thus, an asymptotic 95% confidence interval can be constructed. To calculate the upper 2.5% point of the standard normal distribution, choose Compute from the Transform menu and enter

Z = IDF.NORMAL(0.975,0,1)

The *z* value, found in the Data Editor, is 1.9600. The confidence interval, then, is

Var(leaf) ± 1.9600 × standard error of Var(leaf)

Since the estimate of *Var(leaf)* is 0.05387 and the standard error of *Var(leaf)* is 0.03927, the asymptotic 95% confidence interval for *Var(leaf)* is (–0.02310, 0.1308). Often the interval is reported as (0, 0.1308) because *Var(leaf)* is a non-negative quantity by definition. For more about negative variances, see "Negative Variance Estimates" on p. 181.

Similarly, the asymptotic 95% confidence interval for *Var(Error)* is (0.001319, 0.01188).

You can examine the estimates at various stages of the iteration by requesting the iteration history table. The table helps to identify problems when the iteration fails to converge within the maximum number of iterations or is terminated because iteration cannot be continued. For this example, the iteration history table (Figure 9-8) does not indicate any problems, and the iteration converged at the third iteration with the desired precision.

Figure 9-8
Iteration history

Iteration	Log-likelihood	Var(leaf)	Var(Error)
0	2.292	.074	.060
1	7.544	.013	.007
2	10.429	.054	.007
3	10.429[1]	.054	.007

Dependent Variable: pctca
Method: Maximum Likelihood Estimation

[1]. Convergence achieved.

Negative Variance Estimates

Variances by definition are non-negative quantities. Indeed, you might think that estimates for the variance components are always non-negative; but computationally, the ANOVA method and the minimum normal quadratic unbiased estimation method sometimes produce negative estimates. There is no mechanism in the ANOVA method or in the minimum normal quadratic unbiased estimation method that will prevent negative variance estimates from occurring. However, such occurrences do not imply that the computational algorithms are incorrect.

When a negative variance estimate is obtained, an immediate question is: What does a negative variance estimate indicate? There seems to be no definite answer to this question. Such negative variance estimates are associated with the data or the model. Some possible explanations are:

- The variation of observations may be too large for the sample size, producing negative variance estimates even though the true variances are positive. Try to collect more data in the hope that a larger sample size will yield positive estimates.

- Outliers or erroneously recorded observations are in the data. Identify these observations and handle them appropriately.

- The true value of the variance component may be small or 0. This is usually the case when the negative estimate has a large absolute value. Taking the variance component to be 0 is equivalent to dropping the corresponding random effects from the model.

- The method of estimation is not appropriate. For example, using the ANOVA method on a highly unbalanced data set with empty cells is more likely to produce negative variance estimates. You may want to use the maximum-likelihood or the restricted maximum-likelihood methods.

- The specified model is not correct. The covariance structure of data assumed under the variance components model may not be appropriate for your data. Sometimes a negative variance estimate indicates that observations in your data are negatively correlated.

The above reasons are some of the common possibilities, and they are in no way exhaustive. Users interested in the problem of negative estimates can find more information in LaMotte (1973) and Hocking (1985).

Nested Design Model for Two-Way Classification

A more complicated model can include more than one factor, and the factors can be crossed or nested. Consider the data about test ovens described in Bowker and Lieberman (1972). A quality assurance engineer for an electronics components manufacturing firm claims that the 36 ovens used by this firm for testing the life of the various components are not homogeneous. To determine whether or not the claim can be justified, the engineer conducted an experiment using a single type of electronic component. Three randomly selected ovens were used for the experiment. The electronic component was tested at the two temperatures normally used for life testing of that component. Each component was operated in an oven until it failed. Then the lifetime, in minutes, of the component was recorded. Three components were tested per oven-temperature combination. The original data are shown in Table 9-1.

Table 9-1
Oven temperature data

Temperature	Oven 1	Oven 2	Oven 3
550° F	237, 254, 246	208, 178, 187	192, 186, 183
600° F	178, 179, 183	146, 145, 141	142, 125, 136

Hemmerle and Hartley (1973) deliberately excluded two values (underlined in the table above) in the oven data to introduce a set of unbalanced data. This new set of data is analyzed in this section.

The factor *oven* is clearly a random factor because the three ovens used in the experiment are just a sample of the larger population of 36 ovens. Temperature (*tempture*) is also treated as a random factor because 550° F and 600° F were just two of the many temperatures that could be used for testing. Since there is no information on whether or not the ranges of temperature for testing were the same for all of the ovens, we further assume that *tempture* is nested within *oven*. Hence, a nested design model is used.

The SPSS data file contains four variables: *oven*, *tempture*, *lifetime*, and *qusehh*, with one case for each lifetime recorded. *Lifetime* is the lifetime of a component that is the dependent variable, *oven* is the oven factor with three levels, *tempture* is the temperature factor with two levels, *tempture(oven)* is the temperature within the oven, and *qusehh* is the SPSS weight variable that is used to exclude the two cases not considered in Hemmerle and Hartley (1973) by assigning zero weights to them and unit weights to all other cases. Using these variables, the nested design model is represented as

$$lifetime = \text{Constant} + oven + tempture(oven) + \text{Residual}$$

where both *oven* and *tempture(oven)* are random effects.

SPSS offers several methods to estimate the variance components *Var(oven)*, *Var(tempture(oven))* and *Var(Error)*. However, not all of them are suitable for unbalanced data; for example, the ANOVA Type I method is not suitable. The minimum normal quadratic unbiased estimation (MINQUE) method is used here because it involves no normality assumption. Without any knowledge of what prior values should be used, estimates using both prior values schemes, MINQUE(0) and MINQUE(1), are computed.

To produce the output, open the data *Oven tests.sav* and from the menus choose:

Data
  Weight Cases...

⊙ Weight cases by

  Frequency Variable:
  ▶ qusehh

Analyze
  General Linear Model
    Variance Components...

▶ Dependent Variable: lifetime
▶ Random Factor(s): oven tempture

Model...
  Specify Model
  ⊙ Custom
  Model:
  ▶ oven (R) tempture(R)

Options...
  Method
  ⊙ MINQUE
  Random Effect Priors
  ⊙ Zero

Paste

In the syntax window, at the end of the DESIGN subcommand, type (oven) to specify the nesting effect, tempture(oven). The modified syntax is

```
VARCOMP lifetime BY oven tempture
 /RANDOM = oven tempture
 /METHOD = MINQUE(0)
 /DESIGN = oven tempture(oven)
 /INTERCEPT = INCLUDE.
```

Run the syntax by clicking the Run Current tool. For MINQUE with the other prior values scheme, change the method to

```
 /METHOD = MINQUE(1)
```

and run the syntax again. The results for both methods are shown in Figure 9-9.

Figure 9-9

Variance estimates for the MINQUE(0) and MINQUE(1) methods

Component	Estimate
Var(oven)	85.923
Var(tempture(oven))	1940.664
Var(Error)	-157.119[1]

Dependent Variable: lifetime
Method: Minimum Norm Quadratic Unbiased Estimation
(Weight = 0 for Random Effects, 1 for Residual)

Component	Estimate
Var(oven)	102.839
Var(tempture(oven))	1545.921
Var(Error)	70.067

Dependent Variable: lifetime
Method: Minimum Norm Quadratic Unbiased Estimation
(Weight = 1 for Random Effects and Residual)

[1]. For the ANOVA and MINQUE methods, negative variance component estimates may occur. Some possible reasons for their occurrence are: (a) the specified model is not the correct model, or (b) the true value of the variance equals zero.

Estimates of the residual variance based on the two prior value schemes contradict each other. In order to decide which scheme gives a more reasonable answer, estimates based on another method are computed. The ANOVA Type III method is used, since Type III sums of squares are suitable for unbalanced data. Also, the ANOVA method does not require any normality assumption. The results are shown in Figure 9-10.

Figure 9-10

Variance estimates for the ANOVA Type III method

Component	Estimate
Var(oven)	9.750
Var(tempture(oven))	1475.803
Var(Error)	78.633

Dependent Variable: lifetime
Method: ANOVA (Type III Sum of Squares)

The first finding is that the signs of the MINQUE(1) estimates and the ANOVA Type III estimates do agree. Second, values of the corresponding variance estimates are close except that of *Var(oven)*. Based on these findings, it seems that the second scheme of prior values is more appropriate than the first scheme for this data.

Based on the MINQUE(1) estimates (using the uniform scheme of prior values) in Figure 9-9, the variability among the ovens is about 1.5 times the size of the residual variance, and it contributes

$$102.839/(102.839 + 1545.921 + 70.067) = 5.98\%$$

to the total variance. This result supports the engineer's claim, and efforts should be spent on improving homogeneity of the ovens. Notice that the variance of the temperature-within-oven effect is substantially larger than the residual variance—22 times, to be exact. This effect alone explains

$$1545.921/(102.839 + 1545.921 + 70.067) = 89.94\%$$

of the total variance. This strongly suggests that a very significant and important part of the variance of an electronic component's lifetime is attributed to the temperature used for testing.

Univariate Repeated Measures Analysis Using a Mixed Model Approach

A notable property of longitudinal studies and repeated measures experiments is that each subject is observed at several different times (not necessarily equally spaced) or under different experimental conditions. A classical technique is to apply a univariate mixed model to the data (see Winer, Brown, and Michels, 1991). This model assumes that observations from the same subject have a constant variance and a common correlation. In other words, the variance-covariance matrix of observations from each subject exhibits the **compound symmetry** structure. The common correlation is often called the **intra-class correlation**. In this example, we show how to estimate this intra-class correlation using the Variance Components procedure.

A study was designed to investigate whether boys and girls have different growth rates. Using the distance, in millimeters, from the center of the pituitary to the pteryo-maxillary fissure as a measure of growth, data were collected from each of 11 girls and 16 boys at ages 8, 10, 12, and 14 by investigators at the University of North Carolina Dental School. Occasionally, this distance decreases with age because the distance represents the relative position of two points. The data appeared in Potthoff and Roy (1964) and were again analyzed by Jennrich and Schluchter (1986). The complete data are shown in Figure 9-11.

Figure 9-11
Data for growth study

		Girl				Boy			
		8	10	12	14	8	10	12	14
Distance (mm) from center of pituitary to pteryo-maxillary fissure	1	21.0	20.0	21.5	23.0	26.0	25.0	29.0	31.0
	2	21.0	21.5	24.0	25.5	21.5	22.5	23.0	26.5
	3	20.5	24.0	24.5	26.0	23.0	22.5	24.0	27.5
	4	23.5	24.5	25.0	26.5	25.5	27.5	26.5	27.0
	5	21.5	23.0	22.5	23.5	20.0	23.5	22.5	26.0
	6	20.0	21.0	21.0	22.5	24.5	25.5	27.0	28.5
	7	21.5	22.5	23.0	25.0	22.0	22.0	24.5	26.5
	8	23.0	23.0	23.5	24.0	24.0	21.5	24.5	25.5
	9	20.0	21.0	22.0	21.5	23.0	20.5	31.0	26.0
	10	16.5	19.0	19.0	19.5	27.5	28.0	31.0	31.5
	11	24.5	25.0	28.0	28.0	23.0	23.0	23.5	25.0
	12	.	.	.	.	21.5	23.5	24.0	28.0
	13	.	.	.	.	17.0	24.5	26.0	29.5
	14	.	.	.	.	22.5	25.5	25.5	26.0
	15	.	.	.	.	23.0	24.5	26.0	30.0
	16	.	.	.	.	22.0	21.5	23.5	25.0

You may need to rearrange data so that it is most convenient to specify a univariate mixed model using SPSS. In this example, a categorical variable named *subject* is created to give each individual a unique identification. Girls 1 to 11 in Figure 9-11 are assigned *subject* values 1 to 11. Boys 1 to 16 are assigned values 12 to 27. A second categorical variable, called *age*, is created with four levels that correspond to the four ages (8, 10, 12, and 14). A third categorical variable, called *gender*, uses the characters *F* for female and *M* for male. Finally, *distance* is the dependent variable. There is one case for each distance.

Using repeated measures terminology, *gender* is a between-subjects factor and *age* is a within-subjects factor because each subject is observed at these four ages. Both *gender* and *age* are fixed factors. The variable *subject,* on the other hand, is a random factor because its levels are arbitrarily chosen solely for identification purposes.

The following design uses a common technique (for example, see Winer, Brown, and Michels, 1991) to formulate a repeated measures analysis as a mixed model:

distance = Constant + *gender* + *subject(gender)* + *age* + *age*gender* + Residual with the effect *subject(gender)* being random. The corresponding mathematical model is

$$d_{ijk} = \mu + g_i + s_{k(i)} + a_j + (ag)_{ij} + e_{ijk}$$

$$k = 1, ..., n_i, j = 1, 2, 3, 4 \text{ and } i = 1, 2$$

where $n_1 = 11$; $n_2 = 16$; d_{ijk} is the distance of the kth individual observed for the ith level of *gender* and the jth level of *age*; and g_i is the fixed-effect parameter corresponding to the ith level of *gender*. Similarly, a_j and $(ag)_{ij}$ are the fixed-effect parameters that correspond respectively to the jth level of *age* and the (i,j)th level of the interaction effect *age*gender*. $s_{k(i)}$ is a random-effect parameter that corresponds to the kth individual within the ith level of *gender*, and e_{ijk} is the residual. The usual assumptions are: $s_{k(i)}$ are uncorrelated, each has zero mean and variance σ_s^2, and the residuals are also uncorrelated, while each has zero mean and variance σ_e^2. Furthermore, the random-effect parameters and the residuals are also uncorrelated.

Consider, for example, the distance observed from the first boy at age 10 (that is, d_{221}). It follows from the above distributional assumptions that the variance of this observation is $\sigma_s^2 + \sigma_e^2$. Next, consider the distance observed from the same boy at age 14 (that is, d_{241}). Its variance also equals $\sigma_s^2 + \sigma_e^2$. The covariance between these two distances is σ_s^2 because the two corresponding residuals are uncorrelated. In general, all distances observed from any boy or girl have equal variances, and the common value is $\sigma_s^2 + \sigma_e^2$. Also, the covariance between any two distances observed at two different ages is equal to σ_s^2. Therefore, the correlation between any two distances observed at two different ages is the same. The common correlation value is called the **intra-class correlation**. In this example, it is equal to $\sigma_s^2/(\sigma_s^2 + \sigma_e^2)$. Since both the numerator and the denominator are positive, the intra-class correlation is also positive.

All estimation methods except the maximum-likelihood method give the same variance component estimates. For purposes of illustration, only estimates based on the maximum-likelihood method and those based on the restricted maximum-likelihood method are displayed here. In order to obtain maximum precision, the tables are edited, and up to eight decimal points are shown.

Open the file *Growth study.sav* and from the menus choose:

Analyze
  General Linear Model
    Variance Components...

Dependent Variable: distance
Fixed Factor(s): gender, age
Random Factor(s): subject

Model...
  ⊙ Custom
  Model:
    ▶ gender, age, subject, age*gender

Options...
  Method
  ⊙ Maximum likelihood
  Display
  ☑ Iteration history

Paste

After you paste the syntax, indicate nesting by adding (gender) after subject. The modified syntax is:

```
VARCOMP distance  BY subject    gender age
 /RANDOM = subject
 /METHOD = ML
 /CRITERIA = ITERATE(50)
 /CRITERIA = CONVERGE(1.0E-8)
 /PRINT = HISTORY (1)
 /DESIGN = gender age subject (gender) age*gender
 /INTERCEPT = INCLUDE .
```

After generating the output, double-click the variance estimates table in the Viewer. Highlight the cells containing numbers and change the Cell Properties format to eight decimal places. You may need to drag the right boundary of the cells to display this larger number of decimal places.

Change the method to REML, and run the command again. The variance estimates are shown in Figure 9-12 and Figure 9-13.

Figure 9-12
Variance estimates for ML

Component	Estimate
Var(subject(gender))	3.04202616
Var(Error)	1.82873878

Dependent Variable: distance
Method: Maximum Likelihood Estimation

Figure 9-13
Variance estimates for REML

Component	Estimate
Var(subject(gender))	3.28538826
Var(Error)	1.97503788

Dependent Variable: distance
Method: Restricted Maximum Likelihood Estimation

Based on the maximum-likelihood estimates, the estimate for intra-class correlation is

$$\rho = \frac{3.04202616}{(3.04202616 + 1.82873878)} = 0.62454793$$

whereas the correlation estimate based on the restricted maximum-likelihood estimates is

$$\rho = \frac{3.28538826}{(3.28538826 + 1.97503788)} = 0.62454793$$

Although the variance component estimates are different, the two intra-class correlation estimates are surprisingly close (up to eight decimal points). In order to compute the asymptotic standard errors for these two correlation estimates, you need the asymptotic variance-covariance matrices for the variance component estimates. They are shown in Figure 9-14 and Figure 9-15.

Figure 9-14
Asymptotic covariance matrix for ML

	Var(subject(gender))	Var(Error)
Var(subject(gender))	.91215920	-.02064374
Var(Error)	-.02064374	.08257495

Dependent Variable: distance
Method: Maximum Likelihood Estimation

Figure 9-15
Asymptotic covariance matrix for REML

	Var(subject(gender))	Var(Error)
Var(subject(gender))	1.14905789	-.02600516
Var(Error)	-.02600516	.10402066

Dependent Variable: distance
Method: Restricted Maximum Likelihood Estimation

Using the delta method (see Johnson, Kotz, and Kemp, 1992), the variance of the intra-class correlation is approximated by

$$\text{var}(\rho) = \text{var}(\sigma_s^2/(\sigma_s^2 + \sigma_e^2))$$

$$= (\sigma_e^4\text{var}(\sigma_s^2) + \sigma_s^4\text{var}(\sigma_e^2) - 2\sigma_s^2\sigma_e^2\text{cov}(\sigma_s^2, \sigma_e^2))/(\sigma_s^2 + \sigma_e^2)^4$$

With the maximum-likelihood method,

$$\text{var}(\sigma_s^2) = 0.91215920, \text{var}(\sigma_e^2) = 0.08257495, \text{and}$$

$$\text{cov}(\sigma_s^2, \sigma_e^2) = -0.02064374$$

The approximate variance of the intra-class correlation is

$$\text{var}(\rho) \approx \frac{\left(1.82873878^2 \times 0.91215920 + 3.04202616^2 \times 0.08257495 - 2 \times 3.04202616 \times 1.82873878 \times \left(-0.02064374\right)\right)}{\left(3.04202616 + 1.82873878\right)^4}$$

$$= 0.00718555$$

Taking the square root of $\text{var}(\rho)$ gives the standard error of the intra-class correlation. Based on the maximum-likelihood estimates, the standard error equals 0.08476761.

Since the maximum-likelihood variance estimates are asymptotically normally distributed, it follows that the intra-class correlation is also asymptotically normal. Thus, you can compute an asymptotic 95% confidence interval for this intra-class correlation as $\rho \pm Z_{0.025} \times \text{se}(\rho)$, where Z is the upper 2.5% point of the standard normal distribution. From the Transform menu, choose Compute and enter

$Z_{0.025}$ = IDF.NORMAL(0.975,0,1)

The $Z_{0.025}$ value is 1.95996400. Hence, the asymptotic 95% confidence interval is (0.45840647, 0.79068939).

Similarly, based on the restricted maximum-likelihood variance estimates, the standard error of the intra-class correlation is 0.08809308. The asymptotic 95% confidence interval for the intra-class correlation is (0.45188866, 0.79720720). Although both the maximum-likelihood and the restricted maximum-likelihood methods give very close estimates for the intra-class correlation, the maximum-likelihood method does produce a narrower confidence interval.

For an intra-class correlation estimate based on the ANOVA method, see Searle, Casella, and McCulloch (1992).

Background Information

The following sections include technical background information about variance components models and the types of estimation methods. For readers who are interested in a thorough discussion of this subject, see Rao and Kleffe (1988) and Searle, Casella, and McCulloch (1992).

Model

A variance components model analyzes the contribution of each random effect to the total variation in each cross-classification category. Each cross-classification constitutes a cell and each categorical variable is called a factor. A variance components model formulates the value of the dependent variable in each cell as the sum of a linear combination of parameters and a residual error term. The parameters are identified by association with the categorical variables and the covariates in the model. The parameters are then classified into fixed-effects parameters and random-effects parameters. The classification is based on the nature (fixed or random) of the effects with which the parameters are associated. The residual error term is assumed to have a zero mean and a constant variance across all the cells. The mathematical model for the observed value of the dependent variable in a cell is given by

$$y_{ik} = \mathbf{x}_{io}\beta_o + \sum_{s=1}^{m} \mathbf{x}_{is}\beta_s + \varepsilon_{ik}, \quad k = 1,\ldots, n_i \text{ and } i = 1, \ldots, r$$

where y_{ik} is the kth repetition within the ith cell, $\mathbf{x}_{io}$ is the portion of the ith row of the design matrix associated with the fixed effects, β_o is the vector of fixed-effects parameters, $\mathbf{x}_{is}$ is the portion of the ith row of the design matrix associated with the sth random effect, β_s is the vector of the sth random-effect parameters, ε_{ik} is the residual error associated with the value y_{ik}, n_i is the number of repetitions within the ith cell, m is the number of random effects, and r is the number of cells.

When there are no random effects (that is, $m = 0$), this equation corresponds to a general linear model with fixed effects only.

Distribution Assumptions

Various assumptions are made when you use variance components analysis.

Random Effects

Under a variance components model, the vectors $\beta_1, \ldots, \beta_m$ are random vectors. In other words, elements of these vectors are random variables. Therefore, these random vectors have their expected values and variance-covariance matrices.

For the sth random effect:

- Each element of the vector β_s has zero expected value.

- Variances of the elements of the vector β_s are the same. Denote the common variance as σ_s^2.

- Elements of the vector β_s are uncorrelated. In other words, covariance between *any* two elements is 0.

It is further assumed that the random vectors $\beta_1, \ldots, \beta_m$ are mutually uncorrelated. Thus, the covariance between an element from a random vector and an element from a different random vector is always 0.

Residual Error Term

It is often assumed that:

- The residual error term has zero expectation and constant variance across all cells. Denote the constant variance as σ_ε^2.

- The residuals from different cells are uncorrelated.

- The residual error term is uncorrelated with all elements of the random-effect vectors.

Variance Decomposition

With the above assumptions, the total variance of an observation in the *i*th cell is

$$\text{var}(y_{ik}) = \sum_{s=1}^{m} \sigma_s^2 \|\mathbf{x}_{is}\|^2 + \sigma_\varepsilon^2, \quad k = 1, \ldots, n_i \text{ and } i = 1, \ldots, r$$

where $\|\mathbf{x}_{is}\|^2$ is the sum of squares of all elements of the row vector $\mathbf{x}_{is}$. Hence the data variance is decomposed into a weighted sum of the variances of the random effects and the residual variance. The variance of a random effect is called a **variance component**.

A logical conclusion from the above assumptions is that observations in a variance components model are correlated. This correlation clearly distinguishes between a variance components model and a general linear model. Because the observations are correlated, special estimation methods are used to estimate the variance components.

Estimation Methods

The following four estimation methods are available in the Variance Components procedure:

- ANOVA
- Minimum normal quadratic unbiased estimation (MINQUE)
- Maximum likelihood (ML)
- Restricted maximum likelihood (REML)

ANOVA Method

The ANOVA method first computes sums of squares and expected mean squares for all effects following the general linear model approach. Then a system of linear equations is established by equating the sums of squares of the random effects to their expected mean squares. The variables in the equations are the variance components and the residual variance. Any solution, if one exists, to this system of linear equations constitutes a set of estimates for the variance components.

This method is computationally less laborious, and the estimates are statistically unbiased. However, negative variance estimates can happen, and the variance-covariance matrix of the estimates is difficult to obtain even asymptotically.

The Variance Components procedure offers two types of sums of squares: Type I and Type III. A discussion of choosing the appropriate type can be found in Speed (1979).

MINQUE Method

The MINQUE method requires a set of *a priori* values for the variance components or the ratios of the components to the residual variance. The estimators are then functions of the data and of the prior values. When the prior values are proportional to the true but unknown values of each variance component (or ratios of each component to the residual variance), the estimates achieve minimum variance in the class of all unbiased, translation invariant quadratic estimators. Since the variance components are unknown, the correct prior values are seldom found. Therefore, the estimators are unlikely to possess the above optimal properties. Despite this fact, the MINQUE method is popular because of its considerable flexibility with respect to the form of models that can be fitted. For a summary of the method, see Rao (1973). For an in-depth discussion of the method, see Rao and Kleffe (1988).

The Variance Components procedure offers two schemes of prior values. The first scheme, MINQUE(0), assigns zero prior values to ratios of variance components to the residual variance. The second scheme, MINQUE(1), gives the ratios unit prior values. With either scheme, a system of linear equations is established based on the prior values and the data. The variables are the ratios of the variance components to the residual variance, and the residual variance itself. The system of linear equations is then solved to obtain the MINQUE estimates. Details of the estimation procedure can be found in Giesbrecht (1983).

10

Linear Mixed Models

You've probably already analyzed data using the general linear model framework. The *t* test, analysis of variance and covariance, and regression analysis are all special cases of the general linear model in which you predict the values of a normally distributed dependent variable from a linear combination of independent variables. To use tests based on the general linear model, you have to assume that all observations are independent and have a constant variance. The assumption of independence may not seem to be noteworthy, since you've encountered it many times. In reality, the assumption of independence is often violated.

Correlated observations can arise in many ways, especially when observations cluster within higher-level units, such as schools, hospitals, or regions. You know that students within the same school or patients from the same clinic are probably more similar than students from different schools or patients from different clinics. If you take multiple measurements of the same subject, such as in a repeated measures design, observations from the same subject are also not independent. It's tempting to assume that observations are independent, since that leads to the simplest analyses. What you may not know is that the conclusions you draw from such simplified analyses may be incorrect. If the observations in your sample are not independent, you must incorporate this information into your statistical analysis.

The Linear Mixed Model

The **linear mixed model** is an extension of the general linear model that does not require observations to be independent with constant variance. Models with random effects, hierarchical (multilevel) models, and repeated measures designs can all be analyzed with the Linear Mixed Models procedure.

The General Linear Models (GLM) procedure in SPSS allows specification of random effects but estimates them as if they were fixed. The variance components estimates are calculated from expected mean squares, unlike the estimates calculated in the Linear Mixed Models procedure, which are based on restricted maximum likelihood or maximum likelihood estimation. Repeated measures designs can be analyzed with the GLM Repeated Measures procedure using both univariate and multivariate approaches. However, the mixed models approach is more flexible and allows a large number of covariance structures, inclusion of cases with incomplete data, as well as nonconstant times at which data values are obtained.

In this chapter, the Linear Mixed Models procedure is illustrated with a series of examples. The examples are not standalone but are meant to be read in sequence, since statistical concepts are introduced at various points throughout the chapter. Complete algorithms used in Linear Mixed Models can be accessed from the SPSS Web site at *www.spss.com.* Click Support, click Login to Online Tech Support, and then click Login. If you do not have a user ID and password, you can use Guest for both. Next, click Statistics and then Algorithms.

Example 1: Unconditional random-effects models. Data from the High School and Beyond survey (Raudenbush and Bryk, 2002, and Singer, 1998) are used to examine differences in math achievement scores among 160 schools.

Example 2: Mixed models. A continuation of the previous example that includes gender as a student level factor.

Example 3: Hierarchical models. The High School and Beyond math achievement example continues with the analysis of data at both the student and school levels.

Example 4: Random coefficient models. The example continues with the inclusion of the relationship between student socioeconomic levels and math achievement into the model.

Example 5: School-level and individual-level covariates. A school-level covariate, public versus private schools, is introduced into the model.

Example 6: A three-level hierarchical model. Scores on a smoking knowledge test are analyzed with students clustered in classrooms that are clustered in schools.

Example 7: Repeated measurements. Changes in "opposite" naming scores over time are analyzed.

Example 8: Residual covariance structure. Jaw measurements for children (Pothoff and Roy, 1964) are analyzed. Different residual covariance structures are evaluated.

Background

You're probably familiar with the statistical concept of dependent and independent variables. In many analyses, you study whether there is a relationship between a dependent variable, such as salary or weight, and one or more independent variables, such as education, age, job seniority, dietary habits, and exercise. Regression and analysis-of-variance techniques are commonly used to test hypotheses about the relationship between dependent and independent variables.

Independent variables can be classified as factors or covariates, depending on whether they are categorical variables used to divide the observations into groups (factors) or whether they are interval measurements, such as age or initial score on a test, whose effects must be "controlled for" (covariates). For example, initial weight could be a covariate in an analysis that examines the effectiveness of four weight loss programs. The type of weight loss program is then a factor with four levels or categories. Each level of the factor can have a different effect on the dependent variable.

Fixed Factors and Random Factors

Factors can be further classified as fixed or random. If the four weight loss programs you administer are all of the programs that you are interested in, weight loss program is a **fixed factor**. On the other hand, if each treatment consists of a different dosage of some therapeutic compound, the four treatments may be viewed as a random sample of possible dosages that might be administered. Treatment would then be considered a **random factor**.

For fixed factors, you test hypotheses about the means of the dependent variable for the various levels of the factor. The underlying statistical model postulates a fixed parameter value for each level of the factor. For example, you want to test whether the average weight loss is equal for the four programs. For a random factor, you test hypotheses about the variability of the distribution of the effects. The distribution is assumed to be normal, with a mean of 0 and an unknown variance. From the estimated variance, you can tell how much the random factor contributes to the variability of the dependent variable. If the variance is small, you know that the effects of the different levels of the factor are all close to 0. This is important in determining which random factors are necessary to properly describe the covariance structure of your data. For example, if the city random effect doesn't have a variance that is significantly different from 0, you may be able to assume that observations for people from the same city are independent.

When you design an experiment or study, you are usually interested in the effects of the fixed factors. Often, the random factors are introduced by your sampling methods. For example, if you are studying the length of stay in the hospital after a particular surgical procedure, the patients will be clustered in hospitals. Even though your interest might be how factors such as age, severity of disease, and payment time affect length of stay, you cannot ignore the hospital effect, since lengths of stay for patients in the same hospital are not independent. Certain hospitals are more "efficient" than others. Hospital is a random effect, since you want your conclusions to apply to a broader range of hospitals than those included in your study. By designating hospital as a random effect in your statistical model, you allow lengths of stay for patients in the same hospital to be correlated. Before beginning analysis of any data set, you should identify random factors that may identify correlated groups of observations and include them in the statistical model.

The Mixed Model: Both Random and Fixed Factors

A single study or experiment can have both fixed and random factors. (The resulting statistical model for the data is called a **mixed model**.) For example, if you are comparing the effectiveness of three methods of treating hypertension in males and females, you can randomly assign the treatment to patients treated by different physicians in different clinics. Treatment and gender are fixed factors, since they are all the genders and treatment methods you are interested in, while physicians and clinics are random factors. The particular physicians and clinics selected for inclusion are a random sample from a population of clinics and of physicians within the clinics.

For the fixed factors, you want to test hypotheses about the equality of average blood pressure for males and females, for the treatment methods, and for combinations of gender and treatment. You know that patients treated by the same physician or in the same clinic might behave more similarly than patients treated by different physicians or in different clinics, so you want to test whether the random factors (physicians and clinics) contribute to the observed variability in blood pressure. If they do, you must include the random clinic and physician factors in any comparison of the effects of the fixed factors.

Example 1
Unconditional Random-Effects Models

Everybody worries about the performance of institutions such as schools and hospitals, since few of us escape either unscathed. One of the most basic questions is whether all of the institutions are equally effective. Raudenbush and Bryk (2002) analyze a subset of data from the High School and Beyond survey, a nationally representative sample of United States public and Catholic high schools. The data set contains information on 7,185 students from 160 schools (90 public and 70 Catholic). The number of students per school ranges from 14 to 67, with an average of 45 students per school. (Data can be found at *http://www.ats.ucla.edu/stat/paperexamples/singer/*). This data set is analyzed in considerable detail by Raudenbush and Bryk (2002) and Singer (1998).

Examining Variability among Schools

One of the variables recorded is a math achievement score (*mathach*). How would you go about studying possible differences in math achievement scores between the schools? One of the first ideas that might come to mind is to use one-way analysis-of-variance to test whether the average math achievement scores differ among the schools. There are several problems with this approach. The usual analysis of variance hypothesis tests are for fixed factors. You test whether population means are equal for all categories of the grouping variable. You must have random samples from each of the populations of interest, and you can draw conclusions about only the populations from which you have random samples. The fixed-effects analysis-of-variance model also requires that observations be independent, with a constant variance. You know that students in the same school are probably not independent. Their scores are more likely to be similar than scores of students in different schools because they are taught by the same teachers and live in the same school district.

In this example, school is a random factor. The schools included in your sample are not, in themselves, of interest. They are a carefully constructed sample of high schools in the U.S. You don't really care about these particular 160 schools. Finding that school 23 has a significantly different average score from school 142 is of limited value. You want to draw conclusions about the population of schools based on the 160 schools you have observed. (School would be a fixed factor if you were, for example, studying whether the five high schools in a particular town have the same average math achievement score and you administer the math achievement test to a random sample of students in each of the five schools of interest.) Instead of thinking about school means, it makes more sense to think about how much of the variability in math achievement scores can be attributed to differences between schools.

Examining the Statistical Models

Consider how a simple linear model differs depending on whether school is considered a fixed factor or a random factor.

The linear model for a design with an intercept and a single **fixed factor** is

$$y_{ij} = \mu + \alpha_j + \varepsilon_{ij} \qquad\qquad \text{Equation 10-1}$$

where y_{ij} is the math achievement score for the ith student in the jth school, μ is the overall (grand) mean, α_j is the effect of the jth school, and ε_{ij} is the error associated with the ith student in the jth school. The errors are assumed to be normally and independently distributed, with a fixed variance of σ^2. In Equation 10-1, the math achievement score for student i in school j is the sum of the grand mean, the jth school effect, and an error component. The α is an unknown parameter.

The linear model for a design with an intercept and a single **random factor** is

$$y_{ij} = \mu + a_j + \varepsilon_{ij} \qquad\qquad \text{Equation 10-2}$$

where μ is the overall (grand) mean, a_j is the effect of the jth school, and ε_{ij} is the error term. Equation 10-2 looks very similar to Equation 10-1. In both models, μ is considered a fixed effect. In Equation 10-1, μ represents the population average value for the observed categories of the fixed factor. In this example, it would represent the average for the 160 schools. In Equation 10-2, μ represents the population average value for all levels of the random factor regardless of whether or not they are observed. It represents the population of schools from which the observed sample of 160 schools was selected.

The school effect is designated as α in the first model and a in the second. That's because in the fixed model, α is a parameter, a fixed unknown population value. (Remember, parameters do not have distributions. They are unknown constants. Statistics, which estimate parameters, have distributions.) In the second model, the a_j are not parameters. Instead, they are assumed to be samples from a normal distribution, with a mean of 0 and an unknown variance σ^2_{school}.

The assumptions about the errors, ε_{ij}, are the same in these two models. The errors are assumed to be independently and normally distributed, with a mean of 0 and a constant variance. You'll see later in the chapter that the random effects specify the covariance structure of the data. (**Covariance structure** is a term for the relationship of the observations to each other. The simplest covariance structure is independence of the observations.) So even though the distribution of the error is the same for both models, the random factor introduces covariance among the observations in the same school.

Specifying an Unconditional Random-Effects Model

To fit a model with a fixed intercept and a random school effect using the Linear Mixed Models procedure, choose:

Analyze
  Mixed Models
    Linear...

Specify Subjects and Repeated
Don't specify anything in this dialog box. Click Continue.

Linear Mixed Models
▶ Dependent Variable: mathach
▶ Factor(s): school

Fixed Effects...
☑ Include Intercept

Random Effects...
▼ Covariance Type: Variance Components
Model: school

Statistics...
☑ Parameter estimates
☑ Tests for covariance parameters

There are several menus within the procedure, and each has a well-defined purpose.

Specify Subjects and Repeated. Identify the variables that define the clusters of observations and, in the case of repeated measures data, the variables that define the levels of the repeated measurement. This menu is discussed in more detail later. For a variance component model, you can skip this menu.

Linear Mixed Models. Identify all variables that can be used in a model and define their role as dependent, factor, or covariate. The dependent variable is the variable whose values are predicted. Factors and covariates are used to predict the values of the dependent variable. Factors have a limited number of distinct values that need not have any underlying order. They divide the observations into groups. For fixed factors, a separate parameter estimate is calculated for each factor level. A covariate is a variable with ordered values. A single regression coefficient is calculated for the main effect of a covariate.

Fixed Effects. Specify all terms in the model to be treated as fixed effects. The intercept is included unless the box is unchecked. (As you will see later in some models, the same variable can be listed as both a fixed effect and a random effect.) You can also change the default type sum of squares for the fixed effects. Don't change the default Type 3 sums of squares unless you have a reason for doing so.

Random Effects. Specify all terms in the model to be treated as random effects. A covariance type must be specified for the random effects. Variance component and unstructured are most commonly used. Subject Groupings is an optional specification that defines the clusters for which the random-effects model is estimated.

Estimation. Select the algorithm to be used for estimating the parameters of the model and the criteria to be used to determine convergence. You can select one of two iterative algorithms: maximum likelihood (ML) or restricted maximum likelihood (REML). You must also use this menu to increase the number of iterations or step-halvings if the algorithm does not arrive at a solution.

Statistics. Select the output for fixed effects and random effects.

EM Means. Select the estimated marginal means for display and pairwise comparisons, and select the multiple comparison procedure to be used for pairwise comparisons. The means are estimated from the model.

Save. Select the predicted values, standard errors, and residuals to be saved in the active file.

Figure 10-1 is the summary of the model specified. You should always look at this table to make sure that you have specified the model that you intended. In this example, Intercept is the single fixed effect and School is the single random effect. Intercept has one level, since it is a single parameter; School has 160 levels, since there are 160 schools in the sample. The covariance structure for the random effect is designated as variance components. Three parameters are estimated: the intercept, the variance of the math achievement scores between schools, and the variance of the math achievement scores within schools (the residual variance).

Figure 10-1
Summary of model specification

		Number of Levels	Covariance Structure	Number of Parameters
Fixed Effects	Intercept	1		1
Random Effects	SCHOOL[1]	160	Variance Components	1
Residual				1
Total		161		3

[1]. As of version 11.5, the syntax rules for the RANDOM subcommand have changed. Your command syntax may yield results that differ from those produced by prior versions. If you are using SPSS 11 syntax, please consult the current syntax reference guide for more information.

Goodness-of-Fit Measures

Various measures that can be used to compare the goodness of fit of mixed models are displayed in Figure 10-2. Models with smaller values of the statistics are better. The basic measure is –2 log likelihood or –2 log restricted likelihood, depending on whether you use maximum likelihood (ML) or restricted maximum likelihood (REML) estimation. The other statistics modify the likelihood to penalize for the number of parameters estimated by the model. The most frequently used are

Akaike (AIC) = $-2l + 2d$
Schwarz (BIC) = $-2l + d \log(n)$

where

l = log likelihood (if ML) or log restricted likelihood (if REML)

n = number of cases (if ML) or number of cases minus the number of fixed parameters (if REML)

d = number of fixed-effect parameters plus number of covariance parameters (if ML) or number of covariance parameters (if REML)

Figure 10-2
Information criteria

-2 Restricted Log Likelihood	47116.793
Akaike's Information Criterion (AIC)	47120.793
Hurvich and Tsai's Criterion (AICC)	47120.795
Bozdogan's Criterion (CAIC)	47136.553
Schwarz's Bayesian Criterion (BIC)	47134.553

The information criteria are displayed in smaller-is-better form.

Tests for Fixed Effects

In the Linear Mixed Models procedure, different tables are produced for fixed effects and for random effects. For fixed effects, you get the usual tests for factors and covariates. Notice that the analysis-of-variance table shown in Figure 10-3 does not include sums of squares because the statistical computations are based on likelihood principles. For fixed effects, the denominator degrees of freedom are based on the Satterthwaite method.

Figure 10-3

Tests of fixed effects

Source	Numerator df	Denominator df	F	Sig.
Intercept	1	156.647	2673.663	.000

Figure 10-3 contains statistics for the intercept, the only fixed effect in this example. The F test for the intercept is a test of the null hypothesis that the intercept (the mean of the math achievement scores in the population of schools) is 0. Usually, that's not a hypothesis of any interest. You can obtain the actual estimate of the population intercept by checking parameter estimates in the Linear Mixed Models Statistics dialog box. In Figure 10-4, you see that the estimate of the overall mean is 12.64. That is the value you obtain if you calculate the average score for each school and then average these averages. The parameter estimate divided by its standard error is labeled t in Figure 10-4. (Note that if you square the value of the t statistic, you will obtain the F value shown in Figure 10-3.) The 95% confidence interval for the population mean is from 12.15 to 13.12.

Figure 10-4

Parameter estimates of fixed effects

Parameter	Estimate	Std. Error	df	t	Sig.	95% Confidence Interval	
						Lower Bound	Upper Bound
Intercept	12.636974	.2443936	156.647	51.707	.000	12.1542419	13.1197058

Tests for Random Effects

The table for random effects is shown in Figure 10-5. It contains variance estimates for the two random components: the school effect and the error, or residual term. The variance estimate for math achievement scores for schools is 8.61. That tells you how much average math achievement scores vary in the population of schools. The residual variance tells you how much scores vary within a school. You see that the residual variance is 39.15, four-and-a-half times as large as the variance component for schools. The sum of the two variances is 47.76. That's the total variance of the math achievement scores. About 18% of the total variance can be attributed to differences between schools. This proportion is also known as the **intraclass correlation coefficient**. If the school variance is a small component of the total variance, that means that most of the variability in scores is due to differences between students, not differences between schools. If the intraclass correlation coefficient is close to 1, the differences between scores at different schools are large compared to the differences of the scores within a school. For example, if all students at a school have the same math achievement score but the averages for the schools are not the same, the intraclass correlation coefficient is 1.

Figure 10-5
Covariance parameter estimates

Parameter		Estimate	Std. Error	Wald Z	Sig.	95% Confidence Interval	
						Lower Bound	Upper Bound
Residual		39.148322	.6606447	59.258	.000	37.8746615	40.4648132
SCHOOL	Variance	8.6140247	1.0788036	7.985	.000	6.7391217	11.0105478

Tests for Covariance Parameters

If all schools in the population have similar average math achievement scores, you expect the school variance component to be small. You can test the null hypothesis that the variance components are 0 by selecting Tests for covariance parameters in the Linear Mixed Models Statistics dialog box. This results in the last four columns shown in Figure 10-5. The Wald test is computed by dividing the covariance parameter estimate by its asymptotic standard error. The Wald Z can be unreliable for small samples. A better alternative is to use the likelihood ratio test. (See "Testing Variance Components with the Likelihood Ratio Test" on p. 210). Based on the observed significance level for the Wald Z, you reject the null hypothesis that in the population, both variance components are 0. You are reasonably confident that the average math achievement scores vary among schools and that math achievement scores vary between students in a school. This also tells you that you cannot treat the observations from the same school as independent. Any statistical model that you build must include the school effect.

The model that you just fit is called a **variance components model**, since the variance of the dependent variable is the sum of the variance of each of the random effects (the variance components) and the error term. In a variance components model, all levels of the same random factor are assumed to be independent and have the same variance. For example, all schools are assumed to be independent. In a model with more than one random factor, the random factors are assumed to be independent of one another. The random factors are also assumed to be independent of the error term.

Chapter 10.

Example 2
Adding a Gender Fixed Effect

You saw that over 80% of the observed variability in math achievement scores remains unexplained after the school effect is included in the model. This means that you should look for additional variables that may be associated with math achievement. One possible predictor is gender. Gender is a fixed effect, since you are interested in drawing conclusions only about the two genders included in your sample. There's no larger population of genders of which these two are a sample.

To correctly examine the relationship between gender and math achievement, you must build a model that includes the effects of both gender and school. The fixed-effects model contains the intercept and the gender effect. The random-effects model has three variance components: school, school by gender, and the residual, or error. (The interaction term between a fixed factor and a random factor is considered to be random.)

To fit this model, from the menus choose:

Analyze
  Mixed Models
    Linear...

Specify Subjects and Repeated
Click Continue.

Linear Mixed Models
▶ Dependent Variable: mathach
▶ Factor(s): female, school

Fixed Effects...
☑ Include Intercept
Model: female

Random Effects...
▼ Covariance Type: Variance Components
Model: school, school*female

Statistics...
☑ Parameter estimates
☑ Tests for covariance parameters

The parameter estimates for the fixed effects are shown in Figure 10-6. Based on the observed significance level, you can reject the null hypothesis that there is no relationship between gender and math achievement scores. Boys seem to have significantly higher scores than girls, since female = 0 (male) has a positive parameter estimate which is significantly different from 0. Since you are treating school as a random factor, your conclusion about the gender effect applies to the population of schools, not just to the schools included in the study. If school were a fixed effect, the conclusions about gender would be limited to just those schools included in the study.

Figure 10-6

Parameter estimates for gender

Parameter	Estimate	Std. Error	df	t	Sig.	95% Confidence Interval	
						Lower Bound	Upper Bound
Intercept	11.972955	.2544725	196.750	47.050	.000	11.4711113	12.4747989
[FEMALE=.00]	1.3718963	.1858565	121.118	7.381	.000	1.0039480	1.7398445
[FEMALE=1.00]	0[1]	0	.	.	.	.	.

[1.] This parameter is set to zero because it is redundant.

Now look at the variance component estimates in Figure 10-7. Both of these estimates are now called **conditional estimates**, since their values depend on the fixed effects in the model. The first line, labeled *Residual*, is the estimate of the unexplained variance of the math scores after the gender, school, and gender-by-school effects have been fit. You see that this is the largest of the variance components and is little changed from the previous model that contained only the school effects. The variance component for schools is 7.94. It has decreased somewhat with the addition of the gender effect to the model. Based on the Wald test, it is still significantly different from 0. You didn't really expect possible differences between genders to affect the school variance component much, since most schools contain similar proportions of boys and girls. The variance component for the school-by-gender interaction is very small compared to the other two variance components. This tells you that the relationship between gender and math achievement scores is approximately the same across the population of schools. Based on the Wald test, you cannot reject the null hypothesis that this variance component is 0. This suggests that you can eliminate the school-by-gender effect from the model.

Figure 10-7

Estimates of covariance parameters

Parameter		Estimate	Std. Error	Wald Z	Sig.
Residual		38.73717	.6589493	58.786	.000
SCHOOL	Variance	7.9413290	1.0440371	7.606	.000
SCHOOL * FEMALE	Variance	.3208397	.2770735	1.158	.247

Testing Variance Components with the Likelihood Ratio Test

As previously indicated, the Wald test can be quite unreliable for small samples, especially for testing hypotheses about variance components that are close to 0. A better test that a variance component is 0 is based on looking at the change in –2 log likelihood between two models that differ only by the covariance parameters of interest. For large samples, the change has a chi-square distribution with degrees of freedom equal to the number of parameters that have been dropped. In this example, the degrees of freedom are 1, and the change in –2LL between the model with the interaction and the model without the interaction is 1.69 ($p = 0.53$). Both the likelihood ratio test and the Wald test are conservative when the null hypothesis is on the boundary of the parameter space (that is, testing that a variance component is 0).

You can calculate the observed significance level for the change in –2LL by using the Transform facility, which is accessed from the SPSS Data Editor. From the menus choose:

Transform
  Compute...

Then, in the Compute Variable dialog box, specify:

Prob = SIG.CHISQ(change, df)

where *change* is the difference in –2LL between the two models and *df* is the degrees of freedom. The variable *Prob* (or whatever variable name you give) is the observed significance level. If the observed significance level is less than 0.05, you can reject the null hypothesis that the covariance parameter is 0.

The test for covariance parameters based on the change in –2LL is valid when using either ML or REML estimation. If you are using ML estimation, the change in –2LL can also be used to test that fixed factors have population coefficients that are 0.

Covariance Structure of the Data

In the previous example, you saw that close to 20% of the variability in math achievement scores can be attributed to differences between schools. This tells you that math achievement scores cluster within a school. Students from the same school are more likely to have similar scores than students from different schools. That's hardly surprising. As you know, clustering of individuals and objects has serious ramifications for statistical analyses. Failure to adjust for clustering will cause you to reject the null hypothesis more often than you should. That is, the Type 1 error rate will increase because sampling related units results in less information than sampling independent units. If you are interested in determining the average math achievement values for students in a particular city, you have much less information if you sample 500 students in each of 10 schools than if you randomly sample 5,000 students from the entire city. That's because students in the same school are not independent. You can't just pool all observations together and ignore the fact that students from the same school are not independent.

When you declare a factor as random, you impose a covariance structure on the data. When school is the only random factor, the intraclass correlation coefficient tells you the correlation between pairs of students in the same school. You assume that the correlation between students is the same for all schools. Pairs of students from different schools are assumed to be independent. Their correlation coefficient is 0. The correlation matrix for all pairs of students is assumed to have a particular form called block diagonal. Imagine a matrix that has as many rows and columns as there are students in the data set. For simplicity, assume that students from the same school are next to each other in the matrix. For example, if there are 10 students in the first school, the first 10 rows and columns of the entire matrix represent pairs of students from the first school. Obviously, students are perfectly correlated with themselves, so the matrix has 1s on the diagonal. For all pairs of students from the same school, the correlation coefficient is ρ. For pairs of students from different schools, the correlation coefficient is 0. The correlation matrix is called a **block diagonal matrix**, since the submatrices with non-zero elements cluster around the diagonal. The school variable defines the submatrices, since there is a separate submatrix for students in each school.

If, instead of correlation coefficients, you have variances and covariances in the matrix, you would have the schools variance component in place of the non-zero correlation coefficients and the sum of the schools variance component and the residual error on the diagonal. Such a covariance matrix is said to have a **compound symmetric structure**.

If there is a school-by-gender random effect in the model as well as a school random effect, the variance-covariance structure of the observations is based on a combination of gender and school. Again, pairs of students from different schools are independent and have a correlation of 0. Pairs of students from the same school have one of two possible correlation coefficients, depending on whether they are of the same gender or not. As you increase the number of random effects in a model, the covariance structure becomes more complex.

The Linear Mixed Model

When you test hypotheses about random effects, you are also testing hypotheses about the covariance structure of the data. To understand why, consider the equation for the general linear model and then for the linear mixed model. The fixed-effects linear model is written as

$$y = X\beta + \varepsilon$$

where y is the vector of observed data, β is the unknown vector of fixed-effects parameters, X is the design matrix for the fixed parameters, and ε is a vector of errors assumed to be normally distributed, with a mean of 0 and a variance of σ^2. The general mixed model contains an additional term:

$$y = X\beta + Z\gamma + \varepsilon$$

where Z is the known design matrix, γ is an unknown vector of random-effects parameters, and ε is an unknown vector of random errors.

Both γ and ε are normally distributed and have expected values of 0 and variances of G and R, respectively. The ε and γ are independent. The design matrix Z is set up in the same way as X, the design matrix for the fixed effects, except that the intercept term is usually not included. (If $Z = 0$ and $R = \sigma^2 I$, the mixed model reduces to the standard linear model.) A mixed model has two parts: the fixed model contains the fixed effects, and the random model contains the random effects.

The distribution of a normally distributed random variable is completely specified by its mean and variance. To estimate parameters in the general linear model, using estimation methods that assume normality, it's enough to assume that the errors are independently and normally distributed, with a constant variance, σ^2, since

$$variance\ (y) = \sigma^2.$$

In the general linear mixed model, the variance of y is considerably more complicated:

$$variance\ (y) = ZGZ' + R$$

In order to model data with a mixed model, you must estimate G and R so that the distribution of y is completely specified. One of the simplest structures for the random effects is the variance components structure you previously specified. In a variance component model, the G matrix is diagonal and contains variance components for the random effects in the model. The R matrix is $\sigma^2 I$, where I is an identity matrix. The random menu is used to define G. In the repeated measures section, you will see how to use the Repeated menu to define R.

Failure to Achieve Convergence

Sometimes when running the Linear Mixed Models procedure, you may get the warning message shown in Figure 10-8. It tells you that the criteria for determining that a solution has been found are not met. The estimation algorithm used is iterative, which means that calculations have to be done repeatedly until the predetermined conditions are met. The first step you should take when you get this message is to examine the solution produced. If you see a footnote in any of the tables telling you that a parameter is redundant, that may be the cause of the problem. A parameter is redundant if it occurs twice in the model. That doesn't necessarily mean that you listed the same term twice; it may mean that two terms you thought were different are identical. If you find a redundant parameter, you should eliminate it and then reestimate the model.

If failure to converge is not the result of an easily identifiable redundant parameter, in the Estimation dialog box, select Print iteration history for every n step(s) and rerun the analysis. If a covariance parameter seems to remain constant for all iterations, it is most likely a redundant parameter. To check whether nonconvergence is the result of a bad starting value, increase the number of scoring steps to 5 or so. If convergence is still not achieved, the reason is probably not a bad starting value. Fixed-effect parameter estimates usually converge quickly. If you see erratic behavior in these, the design matrix may be ill-conditioned. In this case, you may want to increase the singularity tolerance value. You can usually achieve convergence by increasing the maximum number of iterations, increasing the maximum step-halvings, or increasing the number of scoring iterations.

Figure 10-8

Warning message: Convergence not achieved

Warnings

Iteration was terminated but convergence has not been achieved. The MIXED procedure continues despite this warning. Subsequent results produced are based on the last iteration. Validity of the model fit is uncertain.

Example 3
Hierarchical Models

One particular type of mixed model that is applied to problems in various disciplines, such as education, medicine, and policy research, is called the **hierarchical model**. (It's also known as the multilevel linear model, random coefficient regression model, and covariance components model.) (See Raudenbush and Bryk, 2002.) Hierarchical models are used when cases are clustered within larger units and information is recorded for both the case and the larger unit. For example, in the High School and Beyond data, you have information for both individual students and schools. Variables such as gender, minority status, and math achievement score are student-level variables. Variables such as type of school (Catholic or private) and size of school are school-level variables. Additionally, individual-level variables can be summarized and analyzed as school-level variables. For example, average math achievement scores and percentage female and percentage minority can be calculated from case-level data and analyzed at the school level. Hierarchical models make use of information at both levels in a single analysis, so you don't have to worry about what the unit of analysis should be. Hierarchical models are not restricted to two levels but can include any number (for example, students within schools within cities within states).

As an example of a hierarchical model in which students are clustered within schools, consider the relationship between math achievement and socioeconomic status (SES) as measured by various household characteristics. In the literature on hierarchical models, the basic unit of analysis is called a level 1 unit and the higher unit is called a level 2 unit. In this example, students are level 1 units and schools are level 2 units. The conceptualization of hierarchical models involves writing three models— one for each of the levels and then a combined model. It is the combined model that you must specify in the Linear Mixed Models procedure and that is the focus in this discussion. For detailed discussion of the individual-level models for these examples, see Raudenbush and Bryk (2002) or Singer (1998).

Introducing a School-Level Covariate

For each student in the High School and Beyond study, there is a value for SES based on parental education, parental occupation, and parental income. For each school, *meanses* is the average of the SES scores for the students in a particular school. To make interpretation of the results easier, the *meanses* scores are standardized to have a mean of 0. To study the relationship between average SES scores for a school and math achievement, you can estimate the following model:

student score = Intercept $+\beta$ (MEANSES) + school random effect + error

The intercept and the regression coefficient for *meanses* are both fixed effects. School and error are both random effects. The school random effect tells you how much school means vary after controlling for *meanses*. The error tells you how much students within a school vary in math achievement. (Remember, all students in the same school have the same average school SES score, so the variability of the students in a school is random error.)

The model is specified in the Linear Mixed Models procedure as follows:

Analyze
  Mixed Models
    Linear...

Specify Subjects and Repeated
▶ Subjects: school

Linear Mixed Models
▶ Dependent Variable: mathach
▶ Covariate(s): meanses

Fixed Effects...
☑ Include Intercept
Model: meanses

Random Effects...
☑ Include Intercept
▼ Covariance Type: Variance Components
Subject Groupings
▶Combinations: school

Statistics...
☑ Parameter estimates
☑ Tests for covariance parameters

The SPSS commands have changed in several ways. First, note that the variable *school* is specified as a subject in the Specify Subjects and Repeated dialog box. The reason for this is to speed computations. Variables in the Subjects list identify clusters of cases that are independent. By moving *school* into the Subjects list, you tell SPSS that students in different schools are independent. You can move more than one variable into the list if combinations of variable values define groups of independent cases. The variable *meanses* is listed as a covariate in the Linear Mixed Models dialog box and as an effect in the Fixed Effects dialog box. The Random Effects dialog box has changed to reflect the fact that *school* has been specified as a subject variable in the Specify Subjects and Repeated dialog box. *School* is moved into the Combinations box. This tells SPSS to treat cases with different values of the school variable as independent. (Since you don't have to use all variables specified in the Specify Subjects and Repeated dialog box for all random models, you must indicate in the Random Effects dialog box which are to be used for that random model.)

The school variable determines the block diagonal structure of the variance-covariance matrix. That is, pairs of students from different schools are assumed to be independent, and pairs of students from the same school are assumed to be correlated. (Another way of saying this is that all effects specified in the Random Effects dialog box are nested within *school*.) Variables in the Subject Groupings list should not be listed as effects in the random model, since they are already included in the model. You can think of them as being "factored out" in the specification. For example, to estimate the school effect, you include the intercept in the random model instead of the school variable. That fits a separate intercept (mean) for each school. (You can get exactly the same output simply by adding *meanses* to the fixed-effects specification in the previous example, in which *school* is specified as a random effect instead of as a subject combinations variable. Computations will take longer, but the results will be the same.)

Estimates for fixed effects are shown in Figure 10-9. The estimated intercept, 12.65, is the same as in the one-way random-effects model, and the estimated slope is 5.86. For every unit change in *meanses*, the math achievement score is predicted to increase by 5.86 points.

Figure 10-9
Estimates of fixed effects

Parameter	Estimate	Std. Error	df	t	Sig.	95% Confidence Interval	
						Lower Bound	Upper Bound
Intercept	12.649435	.1492801	153.743	84.736	.000	12.3545303	12.9443404
MEANSES	5.8635385	.3614580	153.407	16.222	.000	5.1494606	6.5776163

As shown in Figure 10-10, the estimate of the variance of the school random effect, adjusted for *meanses*, is 2.64. The variance between schools is quite a bit smaller than the variance of 8.61 in the unconditional model, the random-effects model without any predictors of math achievement shown in Figure 10-5. However, based on the Wald Z, you can still reject the null hypothesis that there is no significant variation in school means after adjustment for socioeconomic status. You can easily calculate that about 69% of the variance in math achievement between schools is explained by *meanses* $(8.61 - 2.64)/8.61$.

Figure 10-10
Estimates of covariance parameters

Parameter		Estimate	Std. Error	Wald Z	Sig.
Residual		39.15708	.6608016	59.257	.000
Intercept [subject = SCHOOL]	Variance	2.6387080	.4043386	6.526	.000

Example 4
Random-Coefficient Model

In the previous section, you fit a model that looked at the relationship between average socioeconomic status and math achievement scores. You saw that almost 70% of the variability *between* school math achievement scores could be explained by differences in mean socioeconomic status between the schools. Since you have centered socioeconomic status recorded for each of the students (CSES), you can ask how much of the variability *within* a school can be explained by differences in socioeconomic status among students in that school. (The student-level SES is centered to have a mean of 0 in each school so that the parameter estimates are easier to interpret.) To answer this question, you must fit a separate regression model for each school and then look at the variability of the slopes and intercepts. (This means you are fitting 160 regression models, one in each school.) This is sometimes called a **random coefficient regression model**, since you assume that each school has a randomly assigned slope and intercept from a population distribution of possible slopes and intercepts that relate student socioeconomic status to math achievement.

The covariance parameters for the slope and intercept tell you how much the slopes and intercepts vary between schools. The residual error tells you how much students within a school vary about the school's regression line. The tests of the fixed effects for the student-level socioeconomic status tell you if the **average slope** and **average intercept** are significantly different from 0.

To estimate the model, from the menus choose:

Analyze
  Mixed Models
    Linear...

Specify Subjects and Repeated
▶Subjects: school

Linear Mixed Models
▶Dependent Variable: mathach
▶Covariate(s): cses

Fixed Effects...
☑ Include Intercept
Model: cses

Random Effects...
☑ Include Intercept
Model: cses
▼ Covariance Type: Unstructured
Subject Groupings
▶ Combinations: school

Statistics...
☑ Parameter estimates
☑ Tests for covariance parameters
☑ Covariances of random effects

When you have random effects, you must specify their covariance structure (the matrix **G**). In this example, you have to decide what the relationship is between the slope and intercept parameters, the random effects. Since you don't have any information about this relationship, you can select an unstructured covariance matrix that allows the effects to be correlated. This means that you don't impose any structure but use the data to estimate the variances and covariance of the slope and intercept. For regression effects, an unstructured covariance structure is usually selected. (In the variance components model, the random effects were assumed to be independent.)

Figure 10-11 describes the model that you are fitting. The fixed effects are the intercept and the student-level SES. The random effects are the regression intercept and slope. The covariance structure for the coefficients of the random model is unstructured. The total number of parameters to be estimated is six. There are two fixed-effects parameters: the intercept and the slope. There are four covariance parameters: one for the intercept, one for the slope, one for the covariance of the slope and the intercept, and one for the residual. *School* is identified as the subject variable, indicating that a separate model is fit for each level of the school variable.

Figure 10-11
Model description

		Number of Levels	Covariance Structure	Number of Parameters	Subject Variables
Fixed Effects	Intercept	1		1	
	CSES	1		1	
Random Effects	Intercept + CSES[1]	2	Unstructured	3	SCHOOL
Residual				1	
Total		4		6	

[1]. As of version 11.5, the syntax rules for the RANDOM subcommand have changed. Your command syntax may yield results that differ from those produced by prior versions. If you are using SPSS 11 syntax, please consult the current syntax reference guide for more information.

In Figure 10-12, you see that the intercept (the estimated average school mean) is 12.65. The average slope is 2.19. A one-point increase in the student SES score is expected to result in a 2.19-point increase in the math achievement score. Based on the *t* statistic, the slope is significantly different from 0, leading to the conclusion that student SES scores are significantly related to math achievement within schools.

Figure 10-12
Estimates of fixed effects

[handwritten: est. avg school mean]

Parameter	Estimate	Std. Error	df	t	Sig.	95% Confidence Interval	
						Lower Bound	Upper Bound
Intercept	12.649339	.2445133	156.751	51.733	.000	12.1663727	13.1323048
CSES	2.1931921	.1282588	155.218	17.100	.000	1.9398341	2.4465501

[handwritten: slope: 1 pt ↑ in cses = 2.19 ↑ in DV]
[handwritten: CSES predicts DV]

The covariance parameters for the random effects are shown in Figure 10-13. The variance of the student scores in a school, adjusted for the school regression equation, is labeled *Residual*. The residual of 36.7 is only 6% smaller than the residual of 39.14 in Figure 10-5, which was the variability of the students about the school mean without any adjustment for socioeconomic status. Socioeconomic status does not explain much of the variability in math achievement scores *within* a school.

Figure 10-13
Estimates of covariance parameters

Parameter		Estimate	Std. Error	Wald Z	Sig.
Residual		36.70020	.6257440	58.650	.000
Intercept + CSES [subject = SCHOOL]	UN (1,1)	8.6816434	1.0796259	8.041	.000
	UN (2,1)	.0507473	.4063926	.125	.901
	UN (2,2)	.6939945	.2807858	2.472	.013

[handwritten: variability of school intercepts / covariance of slope + intercept / variability of school slopes]

In Figure 10-13, for the entry labeled *UN(1,1)*, 8.68 is the estimate of the variability of the school intercepts. The variability of the school slopes, labeled *UN(2,2)*, is estimated to be 0.69, and the estimated covariance of the slope and intercept, labeled *UN(2,1)*, is 0.051. You can see this more clearly in Figure 10-14, in which the variances and covariances of the random effects are displayed in an easier-to-understand format.

[handwritten: not sign = no r'ship between slope & intercept]

Figure 10-14
Random-effect covariance structure

	Intercept	SCHOOL	CSES	SCHOOL
Intercept	SCHOOL	8.6816434	.0507473
CSES	SCHOOL	.0507473	.6939945

Unstructured

Both the slope and the intercept values vary quite a bit among the schools. Based on the Wald test, you can reject the null hypothesis that the true population variances for the slopes and intercepts are 0. You cannot reject the null hypothesis that the population covariance between the slope and the intercept is 0. There doesn't seem to be a relationship between the slope and the intercept values. You would see a relationship if, for example, schools with larger intercepts (average scores) also had higher slopes.

It's important to remember the distinction between the two previous analyses. In the first example, you were looking at the relationship between average SES in a school and math achievement. You wanted to know if the differences you observed *between* schools could be explained by differences in average SES. In the second analysis, you were looking at the relationship between math achievement scores and SES *within* a school. There can be a strong relationship between the average SES and school-level scores and little or no relationship between the variables within a school. For the 160 schools, imagine a scatterplot of average SES and average math achievement, in which the points cluster tightly about a straight line. This indicates that there is a strong linear relationship between the variables at the school level. Each of the 160 average points is the summary of student-level SES and math achievement scores. Consider a particular school. It's possible that *within* the school, the relationship between the two variables is weak or non-existent. It may be that all students have similar SES scores and similar math achievement scores. Or there may be a relationship between the two variables but a much weaker one than that seen for the school means.

Example 5
A Model with School-Level and Individual-Level Covariates

You have seen that schools differ in average math achievement scores and that there is significant variability in the association between SES and achievement between schools. No doubt, many explanations can be offered for the observed findings. One of the factors that can be considered is whether a school is public or Catholic.

You can include an additional predictor in your model, *sector*, which is coded 0 for public schools and 1 for Catholic schools. This is a variable measured at the school level. Consider a model with the variables *meanses*, *cses*, and *sector* and all of their interactions as the fixed effects and the same random-coefficients model as in the previous example.

To fit this model, from the menus choose:

Analyze
  Mixed Models
    Linear...

Specify Subjects and Repeated
▶ Subjects: school

Linear Mixed Models
▶ Dependent Variable: mathach
▶ Covariate(s): sector, meanses, cses

Fixed Effects...
☑ Include Intercept
Model: meanses, sector, cses, cases*meanses, cses*sector, meanses*sector, cses*meanses*sector

Random Effects...
☑ Include Intercept
▼ Covariance Type: Unstructured
Model: cses
Subject Groupings
▶ Combinations: school

Statistics...
☑ Parameter estimates
☑ Tests for covariance parameters

In the above specification, *sector* is treated as a covariate, since it has only two values, coded as 0 and 1. You can specify the full factorial model without entering each of the terms by choosing the three main effects, selecting Build terms, and then selecting Factorial from the drop-down list. Tests of the fixed effects are shown in Figure 10-15.

Figure 10-15
Estimates of fixed effects in full model

Parameter	Estimate	Std. Error	df	t	Sig.
Intercept	12.18416	.2030354	159.856	60.010	.000
CSES	2.9513307	.1616313	146.180	18.260	.000
MEANSES	5.8731929	.5065303	161.861	11.595	.000
SECTOR	1.2430468	.3051723	147.284	4.073	.000
CSES * MEANSES	1.1288837	.4232154	173.125	2.667	.008
CSES * SECTOR	-1.640651	.2406109	143.068	-6.819	.000
MEANSES * SECTOR	-1.127589	.7355112	148.070	-1.533	.127
CSES * MEANSES * SECTOR	-.1888468	.5996988	159.543	-.315	.753

You cannot reject the null hypothesis that the three-way interaction and the *meanses*-by-*sector* interaction are 0. Once these terms are removed and the model is reestimated, the parameter estimates are displayed in Figure 10-16. The positive coefficient for *sector* tells you that the means for Catholic schools are 1.21 points higher than the means for public schools, controlling for the effect of socioeconomic status. The *meanses* estimate is the school-level slope. The *cses* estimate is the average slope within schools. All effects that involve the variable *cses* are tests for individual school slopes. The interaction of *cses* with *sector* tells you that slopes for individual schools are significantly different in the two sectors. Since the coefficient is negative, you know that the school-level slopes are smaller in Catholic schools. The *cses*-by-*meanses* interaction tells you that the school-level slopes are larger for schools with higher *meanses*.

Figure 10-16
Estimates of fixed effects in reduced model

Parameter	Estimate	Std. Error	df	t	Sig.
Intercept	12.11358	.1988073	159.893	60.931	.000
CSES	2.9387626	.1550929	139.313	18.948	.000
MEANSES	5.3391183	.3692988	150.970	14.457	.000
SECTOR	1.2166722	.3063854	149.600	3.971	.000
CSES * MEANSES	1.0388706	.2989010	160.562	3.476	.001
CSES * SECTOR	-1.642583	.2397914	143.353	-6.850	.000

Estimates of the variability of the random effects for this model are shown in Figure 10-17. The variability of the intercept is only 2.38, compared to 8.68 in the previous model. Introduction of *meanses* and *sector* into the model has greatly reduced the variability of the estimated school means, although based on the Wald statistic, you must still reject the null hypothesis that the population variance is 0. However, you cannot reject the null hypothesis that the variance of the slopes is 0 and that the variance of the covariances of the slopes and intercepts is 0. This means that most of the variability in the slopes between schools has been explained by introduction of the school-level covariates *meanses* and *sector*. It also suggests that a simpler model, without a random coefficient for slope, may fit the data. As in all statistical analysis, you want to select the simplest model that fits the data well.

Figure 10-17

Estimates of covariance patterns

Parameter		Estimate	Std. Error	Wald Z	Sig.
Residual		36.72113	.6261327	58.648	.000
Intercept + CSES [subject = SCHOOL]	UN (1,1)	2.3818588	.3717483	6.407	.000
	UN (2,1)	.1926034	.2045243	.942	.346
	UN (2,2)	.1013797	.2138116	.474	.635

To eliminate the within-school slope from the random-effects model, remove the variable *cses* from the random-effects model. The first part of Figure 10-18 contains the goodness-of-fit statistics for the model with random slopes; the second half contains the same statistics for a model without random slopes. The change in –2 restricted log likelihood between the two models is 1.12, with two degrees of freedom, since both the slope and covariance of the slope and intercept have been dropped from the model. The observed significance level is large ($p = 0.57$), so you can drop the slope without seriously affecting the goodness of fit of the model.

Figure 10-18

Goodness of fit with and without random slopes

-2 Restricted Log Likelihood	46503.668	-2 Restricted Log Likelihood	46504.794
Akaike's Information Criterion (AIC)	46511.668	Akaike's Information Criterion (AIC)	46508.794
Hurvich and Tsai's Criterion (AICC)	46511.673	Hurvich and Tsai's Criterion (AICC)	46508.796
Bozdogan's Criterion (CAIC)	46543.183	Bozdogan's Criterion (CAIC)	46524.552
Schwarz's Bayesian Criterion (BIC)	46539.183	Schwarz's Bayesian Criterion (BIC)	46522.552

The information criteria are displayed in smaller-is-better form.

The information criteria are displayed in smaller-is-better form.

Example 6
A Three-Level Hierarchical Model

The previous example is a two-level hierarchical model, since students (level 1) are clustered within schools (level 2). Now let's look at an experiment in which students (level 1) are clustered in classrooms (level 2), which are clustered in schools (level 3). The Television School and Family Smoking Prevention and Cessation Project study was designed to look at the effects of school-based curricula for smoking prevention and cessation. A subset of data from 28 Los Angeles schools analyzed by Hedeker (1994) is used in the study. Schools were assigned to one of the four experimental conditions, but the programs were delivered to students within classrooms. Scores on a tobacco and health knowledge scale (THKS), which was administered both before and after the intervention, were evaluated.

In this study, schools and classrooms are random factors, while the intervention is a fixed factor. The schools and classrooms included in the study are a sample from a much larger population of schools and classrooms that could have been included. You would like to draw conclusions about the whole population of students and schools. The four interventions studied are not viewed as a sample from the larger population of interventions. They are the interventions of interest. The four interventions are: television only, classroom curriculum only, both classroom curriculum and television, and neither (the control).

Final and preintervention scores are level 1 measurements, since they are measured at the student level. Intervention is a level 3 measurement, since interventions were assigned to schools.

Fitting an Unconditional Model

Let's ignore the intervention and examine the variability in postintervention knowledge scores within schools, within classrooms in schools, and within students within classrooms within schools. From the menus choose:

Analyze
  Mixed Models
    Linear...

Specify Subjects and Repeated
▶ Subjects: schoolid   *company c*

Linear Mixed Models
▶ Dependent Variable: posthks
▶ Factor(s): classid   *inv. #*

Fixed Effects...
☑ Include Intercept

Random Effects...
☑ Include Intercept
▼ Covariance Type: Variance Components
Model: classid
Subject Groupings
▶ Combinations: schoolid

Statistics...
☑ Parameter estimates
☑ Tests for covariance parameters

These selections result in the three variance component estimates shown in Figure 10-19. The residual, 1.72, is the variability of students within classrooms within schools. The variance component for schools is 0.117; the variance component for classrooms within schools is 0.085. Based on the Wald statistic, for each variance estimate, you can reject the null hypothesis that its population value is 0. Remember, if a variance component is significantly different from 0, there is variability between the levels of the random factor. In this example, it means that the scores cluster within classrooms and within schools. If you express each variance component as a percentage of the sum of the three components, you see that 90% of the total variability is attributed to variability within classrooms, 4% to variability among classrooms within schools, and 6% to variability among schools.

Figure 10-19
Covariance parameters for a three-level model

Parameter		Estimate	Std. Error	Wald Z	Sig.
Residual		1.7235903	.0635931	27.103	.000
Intercept [subject = SCHOOLID]	Variance	.1165968	.0480336	2.427	.015
CLASSID [subject = SCHOOLID]	Variance	.0849790	.0338058	2.514	.012

Let's see if you can improve on the unconditional model by including type of intervention (*group*) as a fixed factor and including preintervention score (*prethks*) as a covariate. As in the previous SES and math achievement example, separate slopes and intercepts can be calculated for each classroom for the relationship between postintervention scores and preintervention scores. (For simplicity, ignore the school random factor at this point.) From the menus choose:

Analyze
  Mixed Models
    Linear...

Specify Subjects and Repeated
▶ Subjects: classid

Linear Mixed Models
▶ Dependent Variable: posthks
▶ Factor(s): group
▶ Covariate(s): prethks

Fixed Effects...
☑ Include Intercept
Model: group prethks

Random Effects...
☑ Include Intercept
▼ Covariance Type: Unstructured
Model: prethks
Subject Groupings
▶ Combinations: classid

Statistics...
☑ Parameter estimates
☑ Tests for covariance parameters

The estimates of the covariance parameters are shown in Figure 10-20. Again, *UN(1,1)* corresponds to the intercepts; *UN(2,2)* corresponds to the slopes; and *UN(2,1)* corresponds to the covariances between the slopes and the intercepts. Based on the Wald test, it appears that the slopes and intercepts do not vary significantly among classrooms when the intervention effect is included in the model. This means that the slope and intercept random effects can be eliminated from the model.

Figure 10-20

Covariance parameter estimates

Parameter		Estimate	Std. Error	Wald Z	Sig.
Residual		1.5832253	.0597885	26.480	.000
Intercept + PRETHKS [subject = CLASSID]	UN (1,1)	.0187377	.0592233	.316	.752
	UN (2,1)	.0116614	.0227970	.512	.609
	UN (2,2)	.0072575	.0109059	.665	.506

(handwritten annotations): ← intercepts don't vary · slopes don't vary

Now let's fit the three-level model with fixed effects for *group* and *prethks* and random effects for both classroom-within-school and school. From the menus choose:

Analyze
  Mixed Models
    Linear...

Specify Subjects and Repeated
▶ Subjects: schoolid

Linear Mixed Models
▶ Dependent Variable: posthks
▶ Factor(s): group classid
▶ Covariate(s): prethks

Fixed Effects...
☑ Include Intercept
Model: group prethks

Random Effects...
☑ Include Intercept
▼ Covariance Type: Variance Components
Model: classid
Subject Groupings
▶ Combinations: schoolid

Statistics...
☑ Parameter estimates
☑ Tests for covariance parameters

EM Means...
▶ Display Means for: group
☑ Compare main effects
▼ Confidence Interval Adjustment: Sidak

Covariance parameter estimates for the three-level model are shown in Figure 10-21. Inclusion of the intervention effect and the covariate effect reduced all of the estimates from those estimates shown in Figure 10-19. Based on the Wald test, you can no longer reject the null hypothesis that the variability associated with schools is 0. However, if you calculate the more reliable likelihood ratio test, you find that the change in –2 restricted log likelihood is 4.5 and the observed significance level is 0.03, indicating that you must reject the null hypothesis that the covariance parameter is 0.

Figure 10-21
Covariance parameters for a three-level model

Parameter		Estimate	Std. Error	Wald Z	Sig.
Residual		1.6022928	.0591287	27.098	.000
Intercept [subject = SCHOOLID]	Variance	.0386376	.0253469	1.524	.127
CLASSID [subject = SCHOOLID]	Variance	.0646646	.0285530	2.265	.024

From the tests of fixed effects shown in Figure 10-22, you see that intervention and group are significantly associated with the final score.

Figure 10-22
Tests of fixed effects

Source	Numerator df	Denominator df	F	Sig.
Intercept	1	72.225	687.425	.000
GROUP	3	21.472	6.611	.002
PRETHKS	1	1587.783	139.099	.000

Estimated means at the average value of the covariates are shown in Figure 10-23. The control group has the smallest mean; the classroom-curriculum-only group has the largest mean.

Figure 10-23
Estimated marginal means

GROUP	Mean	Std. Error	df	95% Confidence Interval	
				Lower Bound	Upper Bound
tv only	2.516[1]	.110	20.849	2.288	2.744
cc only	2.975[1]	.115	22.370	2.737	3.214
both cc and tv	2.826[1]	.112	22.149	2.595	3.058
control	2.334[1]	.113	20.594	2.100	2.568

[1]. Covariates appearing in the model are evaluated at the following values: PRETHKS = 2.07.

Pairwise differences between groups are shown in Figure 10-24. Only the control group is significantly different from both the classroom-curriculum-only group and the combined classroom-curriculum-and-TV group, using the Sidak multiple-comparison procedure.

Figure 10-24

Pairwise comparisons using Sidak multiple comparisons

(I) GROUP	(J) GROUP	Mean Difference (I-J)	Std. Error	df	Sig.[1]	95% Confidence Interval for Difference	
						Lower Bound	Upper Bound
tv only	cc only	-.459	.159	21.627	.051	-.920	.001
	both cc and tv	-.310	.157	21.490	.312	-.764	.143
	control	.182	.157	20.721	.836	-.275	.639
cc only	tv only	.459	.159	21.627	.051	-.001	.920
	both cc and tv	.149	.160	22.249	.933	-.314	.612
	control	.641*	.161	21.485	.004	.175	1.107
both cc and tv	tv only	.310	.157	21.490	.312	-.143	.764
	cc only	-.149	.160	22.249	.933	-.612	.314
	control	.492*	.159	21.361	.031	.033	.952
control	tv only	-.182	.157	20.721	.836	-.639	.275
	cc only	-.641*	.161	21.485	.004	-1.107	-.175
	both cc and tv	-.492*	.159	21.361	.031	-.952	-.033

Based on estimated marginal means

*. The mean difference is significant at the .05 level.

[1]. Adjustment for multiple comparisons: Sidak.

Example 7
Repeated Measurements

In all of the previous examples, each subject had only one value for the dependent variable. You analyzed a single math achievement score in high school and a single tobacco knowledge score immediately after completion of an intervention. In many studies, subjects are observed over time, and the same variable is recorded on more than one occasion. Studies in which multiple measurements of the same variables are made at different points in time or in space are called repeated measurement or longitudinal studies. Special statistical models are needed for analysis of such data, since multiple observations from the same subject are not independent.

Repeated measures designs can be analyzed within the mixed models framework. Instead of viewing students or patients as clustering in classrooms or hospitals, you consider all of the observations from a single subject as a cluster. The level 1 model describes the data for an individual, and the level 2 model describes differences between individuals. (In the previous hierarchical models, all level 1 units (students) were pooled to describe the relationship of variables within a school. The multiple observations for each subject will now substitute for an entire school's observations.) The subjects, like the schools in the previous examples, are considered to be random factors.

Consider data from an experiment described by Willett (1988), Singer (1998), and Willett and Singer (2003). For each of 35 people, you have an "opposite naming" score obtained on each of four days, spaced exactly one week apart. At the first time point, you also have a general cognitive skill (COG) score for each person. These data can be arranged in one of two formats in a data file. In the first format, the subject is the case, and the multiple observations for a subject are given different variable names.
Table 10-1 shows five such records, with possible variable names listed as column headers. Note that if a subject doesn't have measurements for a particular time point, missing values must be specified for the corresponding variable. In this example, the times are the same for all subjects, but that need not be true. Subjects can have different values for the time variables, but in this format, the same number of times must be recorded for each subject, even if some are missing. A different variable name must be given to each measurement and each time.

Table 10-1

Sample from Willet and Singer data

Person	COG	Time1	Score1	Time2	Score2	Time3	Score3	Time4	Score4
1	137	0	205	1	217	2	268	3	302
2	123	0	219	1	243	2	279	3	302
3	129	0	142	1	212	2	250	3	289
4	125	0	206	1	230	2	248	3	273
5	81	0	190	1	220	2	229	3	220

In the second format, each time point for each subject is a case. Consider the data in Table 10-2, in which the first two subjects from Table 10-1 are rearranged into the time-as-case format. Each case in Table 10-1 is converted into four cases, one for each time point. The person identifier and the value of any covariates that are measured only once are repeated for each time point. Unlike the subject format, only the time points at which data are available are included. For example, if person 1 did not have a value at time 3, the entire row for time 3 could be omitted.

Table 10-2
Restructured data

Person	COG	Time	Score
1	137	0	205
1	137	1	217
1	137	2	268
1	137	3	302
2	123	0	219
2	123	1	243
2	123	2	279
2	123	3	302

Converting between Formats

The Linear Mixed Models procedure requires that data be in the time-as-a-case format. The Data Wizard in SPSS can be used to convert data files from one format to another. From the menus choose:

Data
  Restructure...

To convert a data file that has subjects as cases into a data file with time periods as cases, select Restructure selected variables into cases, and then follow the instructions in the dialog boxes.

A Simple Model

To determine whether the mean opposite naming scores are the same for the four time periods, you can estimate the model

Score = Grand mean + Time effect + Subject effect + Error

The grand mean (intercept) and time are fixed factors; subject and error are random factors.

From the menus choose:

Analyze
  Mixed Models
    Linear...

Specify Subjects and Repeated
▶ Subjects: id

Linear Mixed Models
▶ Dependent Variable: score
▶ Factor(s): time

Fixed Effects...
☑ Include Intercept
Model: time

Random Effects...
☑ Include Intercept
▼ Covariance Type: Variance Components
Subject Groupings
▶ Combinations: id

Statistics...
☑ Parameter estimates
☑ Tests for covariance parameters

The tests for the fixed effects are shown in Figure 10-25. You can reject the null hypothesis that the four time means are equal.

Figure 10-25
Tests for fixed effects of time

Source	Numerator df	Denominator df	F	Sig.
Intercept	1	34	1470.004	.000
TIME	3	102.000	111.097	.000

In Figure 10-26, you see that the variance component between subjects is 903. The residual (within-subjects) variance component is 382. You can reject the null hypothesis that the variance components are 0. Of the total variance of 1285, 70% is attributable to variability between subjects.

Figure 10-26
Covariance parameters

Parameter		Estimate	Std. Error	Wald Z	Sig.
Residual		382.0235	53.49399	7.141	.000
Intercept [subject = ID]	Variance	903.2748	242.6088	3.723	.000

Fitting an Unconditional Linear Model

Figure 10-27 is a plot of the values of time and score for the first five cases. Score increases with time, and it appears that a straight line is a reasonable summary of the relationship between score and time.

Figure 10-27
Plot of time and score for five cases

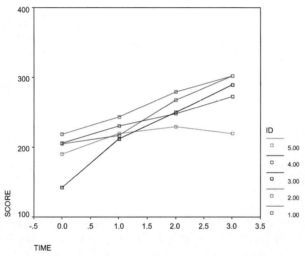

In the school example, you fit a linear model that related socioeconomic status and math achievement. It was called a random coefficients model, since you estimated the intercept and slope separately for each of the schools. In a repeated measures design, you can estimate separate models for each of the subjects in the study. That is, you can estimate the coefficients for the linear regression equation

Score = Intercept + B(Time)

for each of the subjects. Then you can see how much the slopes and intercepts vary among the subjects. These terms, together with the error, are the random effects. You can also test whether the average intercept and slope across persons are different from 0. These are the fixed effects. The variable *time* is no longer a factor variable with four arbitrary levels. Instead, it is a covariate with numeric values 0, 1, 2, 3, and 4.

From the menus choose:

Analyze
  Mixed Models
    Linear...

Specify Subjects and Repeated
▶ Subjects: id

Linear Mixed Models
▶ Dependent Variable: score
▶ Covariate(s): time

Fixed Effects...
☑ Include Intercept
Model: time

Random Effects...
☑ Include Intercept
▼ Covariance Type: Unstructured
Model: time
Subject Groupings
▶ Combinations: id

Statistics...
☑ Parameter estimates
☑ Tests for covariance parameters

Note that the variable *time* is no longer treated as a fixed effect but as a covariate with a meaningful scale. From the fixed-effects estimates in Figure 10-28, you see that the average intercept across subjects is 164.4 and the average slope is 26.96. Starting with an average score of 164, subjects gain 27 points each time they are tested. (The first value of time is 0, so the intercept is the first score.) You can reject the null hypothesis that both the slope and intercept are 0.

Figure 10-28
Estimates of fixed effects

Parameter	Estimate	Std. Error	df	t	Sig.	95% Confidence Interval	
						Lower Bound	Upper Bound
Intercept	164.37429	6.1188486	34	26.864	.000	151.9392892	176.8092822
TIME	26.960000	2.1666037	34	12.443	.000	22.5569316	31.3630684

From the estimates of covariance parameters shown in Figure 10-29, you see that there is significant variation between subjects in both the slope and the intercept. The residual error (the estimated variability between times within a person) decreased from 341 to 159 because the regression equation accounted for part of the observed variability of the scores.

Figure 10-29
Estimates of covariance parameters

Parameter		Estimate	Std. Error	Wald Z	Sig.	95% Confidence Interval	
						Lower Bound	Upper Bound
Residual		159.4771	26.95656	5.916	.000	114.5035958	222.1149381
Intercept + TIME [subject = ID]	UN (1,1)	1198.777	318.3810	3.765	.000	712.3103292	2017.471505
	UN (2,1)	-179.2556	88.96342	-2.015	.044	-353.6206548	-4.8904712
	UN (2,2)	132.4006	40.21070	3.293	.001	73.0089022	240.1064908

Introducing a Covariate

A possible explanation for the observed variability in slopes and intercepts is differences in cognitive ability between subjects. The baseline cognitive score (*COG*) can be introduced as a covariate. Since the scale was administered once for each subject, it can't explain any of the within-subject variability, but it can explain between-subject variability.

From the menus choose:

Analyze
  Mixed Models
    Linear...

Specify Subjects and Repeated
▶ Subjects: id

Linear Mixed Models
▶ Dependent Variable: score
▶ Covariate(s): time, cog

Fixed Effects...
☑ Include Intercept
Model: time cog, time*cog

Random Effects...
☑ Include Intercept
▼ Covariance Type: Unstructured
Model: time
Subject Groupings
▶ Combinations: id

Statistics...
☑ Parameter estimates
☑ Tests for covariance parameters

In Figure 10-30, you see that the parameter estimate for the baseline covariate is not significant, although its interaction with time is significant, indicating that the rate of change is higher for people with higher scores on the baseline covariate.

Figure 10-30
Tests for fixed effects

Source	Numerator df	Denominator df	F	Sig.
Intercept	1	33.000	9.497	.004
TIME	1	33	1.437	.239
COG	1	33.000	.051	.823
TIME * COG	1	33	7.146	.012

The variance components for the intercept and the slope shown in Figure 10-31 are still significantly different from 0. The within-person residual remains at 159, since inclusion of a covariate that does not change with time for an individual cannot affect how well the within-person model fits.

Figure 10-31
Estimates of covariance parameters

Parameter		Estimate	Std. Error	Wald Z	Sig.	95% Confidence Interval	
						Lower Bound	Upper Bound
Residual		159.4771	26.95656	5.916	.000	114.5035958	222.1149381
Intercept + TIME [subject = ID]	UN (1,1)	1236.413	332.4022	3.720	.000	730.0002411	2094.131334
	UN (2,1)	-178.2332	85.42978	-2.086	.037	-345.6725348	-10.7939596
	UN (2,2)	107.2492	34.67670	3.093	.002	56.9084074	202.1210841

The Within-Persons Variance-Covariance Matrix

Consider the simple model you used to test the null hypothesis that the population means at all four times were equal. There is one random effect, the between-subjects error, and the residual is the within-subject error. From the estimated variances of these two effects, you can estimate the variance-covariance matrix of the residuals for pairs of observations from the same subject. For each subject, the variance of the residuals at each of the four time points is estimated to be 1285.30 (the sum of the between-subjects variance and the residual variance). The covariance between the residuals at any two different time points for the same subject is estimated as 903.27. (These variances are shown in Figure 10-26. The residual covariance matrix is shown in Figure 10-32.) You will recognize this as a compound symmetric covariance matrix, since it has one value for all of the diagonal elements and another value for all of the off-diagonal elements. All subjects are assumed to have the same residual covariance matrix.

Although you might initially think that one covariance matrix is as good as another, that's not really true. In many situations, you expect that observations that are closer in time should be more closely related than observations further apart in time. For example, if there is a relationship between time and score, the opposite naming score at time 0 should be more closely related to the score at time 1 than to the score at time 3.

When analyzing repeated measures designs in a mixed model framework, you can directly specify the structure of the residual covariance matrix. If there are no random effects in the model, the residual covariance matrix alone determines the variance of y. You want to find a covariance structure that fits the observed data well and requires estimation of the smallest number of parameters. It is important to identify the correct variance-covariance structure, since the tests of the fixed effects depend on the structure selected.

Specifying the Residual Covariance Structure

When you specify random effects in the Random Model dialog box, they describe the error-covariance structure of the data. You saw that a random-intercepts model results in a compound symmetric structure for the residual covariance matrix. The random-intercepts-and-slopes model with time results in an error-covariance structure that may not be appropriate in many situations, since it increases parabolically over time from a single minimum.

The within-subjects covariance matrix (**R**) can be selected directly using the Repeated specification in the Linear Mixed Models Specify Subjects and Repeated dialog box. Move the variable that identifies the independent subjects into the Subjects list. Cases with different values of the subject variable are assumed to be independent. The covariance between two independent subjects is 0. Then move the variable or variables that identify the repeated observations into the Repeated list. Select the type of repeated covariate matrix from the Repeated Covariance Type drop-down list. For example, you can replicate the analysis shown in Figure 10-25 and Figure 10-26 by specifying directly a compound symmetry model for the variance-covariance matrix of the residuals, as follows.

From the menus choose:

Analyze
  Mixed Models
    Linear...

Specify Subjects and Repeated
▶ Subjects: id
▶ Repeated: time
▼ Repeated Covariance Type: Compound Symmetry

Linear Mixed Models
▶ Dependent Variable: score
▶ Factor(s): time

Fixed Effects...
☑ Include Intercept
Model: time

Statistics...
☑ Parameter estimates
☑ Tests for covariance parameters
☑ Covariances of residuals

The estimated covariances of the residuals are shown in Figure 10-32. This is the variance of the dependent variable. (If you specify the model as a random-effects model, you can't obtain this matrix directly, since you are specifying the **G** and **R** matrices individually.) You see that the same model can be specified in different ways, since you can specify a random-effect covariance structure **G** with the Random specification, or you can specify the error-covariance matrix structure **R** directly with the Repeated specification. You can also use the Repeated and Random specifications together to specify both the **G** and **R** matrices.

Figure 10-32
Residual covariance matrix

	[TIME = 0]	[TIME = 1]	[TIME = 2]	[TIME = 3]
[TIME = 0]	1285.2983	903.27479	903.27479	903.27479
[TIME = 1]	903.27479	1285.2983	903.27479	903.27479
[TIME = 2]	903.27479	903.27479	1285.2983	903.27479
[TIME = 3]	903.27479	903.27479	903.27479	1285.2983

Compound Symmetry

Example 8
Selecting a Residual Covariance Structure

To see what is involved in selecting a residual covariance structure, consider data reported by Pothoff and Roy (1964). A distance measurement from the pituitary gland to a well-defined location in the jaw was recorded for 11 girls and 16 boys at ages 8, 10, 12, and 14. To simplify selection of an appropriate form of the residual covariance matrix, fit a factorial model with fixed effects only and an unstructured covariance matrix, which is the most general form possible.

From the menus choose:

Analyze
　Mixed Models
　　Linear...

Specify Subjects and Repeated
▶ Subjects: id
▶ Repeated: age
▼ Repeated Covariance Type: Unstructured

Linear Mixed Models
▶ Dependent Variable: distance
▶ Factor(s): gender age

Fixed Effects...
☑ Include Intercept
Model: age, gender, age∗gender

Statistics...
☑ Parameter estimates
☑ Covariances of residuals

Note that there are no random effects. The repeated specification is used instead of the random model specification to indicate the covariance structure of the residuals.

Unstructured

The unstructured covariance matrix of the residuals is shown in Figure 10-33. Like all covariance matrices, the matrix is symmetric, since the covariance between the measurement at age 8 and at age 10 is the same as the covariance between age 10 and age 8. On the diagonal are the variances at each of the ages. Unlike the compound symmetry matrix, you don't see a pattern in the variances and covariances. When you fit an unstructured covariance matrix, you estimate each of the coefficients separately.

240

Chapter 10

Figure 10-33
Unstructured residual covariance matrix

	[AGE = 8]	[AGE = 10]	[AGE = 12]	[AGE = 14]
[AGE = 8]	5.4154545	2.7168182	3.9102273	2.7102273
[AGE = 10]	2.7168182	4.1847727	2.9271591	3.3171591
[AGE = 12]	3.9102273	2.9271591	6.4557386	4.1307386
[AGE = 14]	2.7102273	3.3171591	4.1307386	4.9857386

Unstructured

You should examine the unstructured matrix carefully to detect any patterns that might help you to select a simpler covariance structure, since you don't want to estimate any more parameters than are needed to adequately represent the data. From the diagonal elements, you see that the variance is not increasing with age. The variance at age 14 is actually less than the variance at age 8. You also see that pairs of ages that are closer in value don't have larger covariances than pairs that are further apart.

Compound Symmetry

Visual inspection of the unstructured matrix suggests that the residual covariance matrix might be described as compound symmetric. That means that all diagonal variances are equal and all off-diagonal variances are equal. As you saw, that's the covariance structure when you fit a random intercepts model. Figure 10-34 is the residual covariance matrix when compound symmetry is specified as the repeated covariance type. From the compound symmetric variance-covariance matrix, you can easily calculate the correlation coefficient between all pairs of measurements on the same individual. This coefficient, called the intraclass correlation, can be calculated as the ratio of the covariance to the variance. In this example, it is $3.28/5.26 = 0.62$.

Figure 10-34
Compound symmetry residual covariance matrix

	[AGE = 8]	[AGE = 10]	[AGE = 12]	[AGE = 14]
[AGE = 8]	5.2604261	3.2853883	3.2853883	3.2853883
[AGE = 10]	3.2853883	5.2604261	3.2853883	3.2853883
[AGE = 12]	3.2853883	3.2853883	5.2604261	3.2853883
[AGE = 14]	3.2853883	3.2853883	3.2853883	5.2604261

Compound Symmetry

Autoregressive Models

If successive measurements are correlated, the residual covariance matrix might be described by an autoregressive model in which the correlation between observations that are k time periods apart is taken to be $\mathbf{rho}^k$, where *rho* is the intraclass correlation coefficient. Figure 10-35 shows the estimates of the covariance parameters when AR(1) is selected as the repeated covariance type. The estimate labeled *AR1 diagonal* is the estimated variance, and *AR1 rho* is the intraclass correlation coefficient.

Figure 10-35
AR(1) parameter estimates

Parameter		Estimate	Std. Error
Repeated Measures	AR1 diagonal	5.2464580	.9660890
	AR1 rho	.6152662	.0794112

In the residual covariance ($\mathbf{R}$) matrix, shown in Figure 10-36, you see that all variances are estimated to be 5.25. The covariance between measurements for age 8 and age 10 is 3.23 $(5.25*0.62)$, and the covariance between age 8 and age 12 is 1.99 $(5.25*0.62^2)$. As the difference between the ages increases, the covariance decreases. To obtain this output, change the specification of the repeated covariance type to AR(1).

Figure 10-36
AR(1) residual covariance matrix

	[AGE = 8]	[AGE = 10]	[AGE = 12]	[AGE = 14]
[AGE = 8]	5.2464580	3.2279685	1.9860600	1.2219557
[AGE = 10]	3.2279685	5.2464580	3.2279685	1.9860600
[AGE = 12]	1.9860600	3.2279685	5.2464580	3.2279685
[AGE = 14]	1.2219557	1.9860600	3.2279685	5.2464580

First-Order Autoregressive

The first-order autoregressive model is a special case of a more general model, called the **Toeplitz model**. In the Toeplitz model, the covariance depends only on the lag between observations, but the covariances are not constrained to be multiples of the same intraclass correlation coefficient. Separate coefficients are estimated for each lag. The variances are the same for all time points. For four time points, four parameters must be estimated: the variance and the multipliers for lag1, lag2, and lag3. Both the first-order autoregressive and Toeplitz structures should be considered only when there is equal spacing between the time points.

Comparing Covariance Structures

There are a large number of covariance structures that can be used to model the residual matrix. How do you choose among them? One possible tactic is to see how well the overall model fits when a particular structure is used. Table 3 shows the values of three commonly used goodness-of-fit statistics. They are all based on the log likelihood and penalize in different ways for the number of parameters estimated. The −2 log likelihood does not penalize for the number of parameters estimated, and BIC imposes a more severe penalty than does AIC. Remember, the more parameters you estimate, the better your model will fit the data.

For the definitions used by SPSS, see the SPSS Web site at *www.spss.com*. Select Support and then Login to Support. If you do not have a user ID and password, you can use Guest for both. Next, select Statistics and then Algorithms. SPSS calculates these measures so that smaller values are better.

Table 10-3
Comparison of covariance models

Covariance Structure	Covariance Parameters Estimated	−2LL	AIC	BIC
Unstructured	10	414.03	434.04	460.09
Compound symmetry	2	423.41	427.41	432.62
AR(1)	2	434.55	438.55	443.76
Toeplitz	4	418.95	426.95	437.37

Since −2 log likelihood does not penalize for the number of parameters estimated, it will always be smallest for the unstructured covariance matrix, since it requires the largest number of parameters to be estimated. However, it serves as a convenient base from which other models can be evaluated. The compound symmetry and AR(1) models both require two covariance parameters to be estimated; the Toeplitz model requires four. Based on the Akaike information criterion (AIC), the compound symmetry and Toeplitz models are the best. The Bayesian information criterion (BIC), which penalizes the estimation of the additional Toeplitz parameters more severely than does the AIC, suggests that the compound symmetry model fits better than the Toeplitz model.

Comparing Models with a Likelihood Ratio Test

For two models with the same fixed effects, you can test whether there is a significant change in the log likelihood with different covariance structures, provided the covariance models are nested. This means that the covariance structure with fewer parameters must be expressible as a special case of the covariance structure with more parameters. For example, the AR(1) model and the compound symmetry models are special cases of the Toeplitz model. The change in –2LL between Toeplitz and AR(1) is 15.6 with 2 degrees of freedom. The observed significance level is less than 0.001. Comparing compound symmetry and Toeplitz, the change in –2LL is 4.46 with 2 degrees of freedom. The observed significance is $p = 0.11$, indicating that compound symmetry may be the best covariance structure for these data.

Tests of Fixed Effects

In most experimental designs, your primary interest is in tests of the fixed effects. The covariance structure usually does not have much of an effect on the parameter estimates for the fixed effects. It does, however, affect the standard error of the estimate and the observed significance level for hypothesis tests. Table 10-4 contains the F tests and observed significance levels for the fixed effects with different covariance structures. In this example, the hypothesis testing is not much affected by the different structures. Note that the estimates of the fixed effects are unbiased regardless of the error structure selected for the random effects. If you decide to eliminate some of the fixed effects from your model, you should reevaluate the appropriateness of the selected covariance structure for the new model.

Table 10-4
Comparison of tests of fixed effects

	Unstructured	Compound Symmetry	AR(1)	Toeplitz
Age	34.45 (p<0.0005)	35.35 (p<0.0005)	15.48 (p<0.0005)	29.97 (p<0.0005)
Gender	9.29 (0.005)	9.29 (0.005)	11.07 (0.002)	9.19 (0.006)
Age by Gender	2.93 (0.053)	2.36 (0.078)	1.125 (0.345)	2.34 (0.086)

Specifying a Random-Intercepts-and-Coefficients Model

Since *age* is an interval variable, you can also consider a model in which *age* is treated as a covariate instead of as a fixed factor with four levels. You can specify the model in several ways. If you want to specify a random-intercepts-and-coefficients model, from the menus choose:

Analyze
  Mixed Models
    Linear...

Specify Subjects and Repeated
▶ Subjects: id

Linear Mixed Models
▶ Dependent Variable: distance
▶ Factor(s): gender
▶ Covariate(s): age

Fixed Effects...
☑ Include Intercept
Model: gender, age, age*gender

Random Effects...
☑ Include Intercept
▼ Covariance Type: Unstructured
Model: age
Subject Groupings
▶ Combinations: id

Statistics...
☑ Parameter estimates
☑ Tests for covariance parameters
☑ Covariances of residuals

The output of the model is shown in Figure 10-37.

Figure 10-37
Random-intercepts-and-coefficients model estimates

Parameter		Estimate	Std. Error	Wald Z	Sig.	95% Confidence Interval	
						Lower Bound	Upper Bound
Residual		1.7162037	.3302836	5.196	.000	1.1769419	2.5025493
Intercept + AGE [subject = ID]	UN (1,1)	5.7864331	5.1352246	1.127	.260	1.0162583	32.9471438
	UN (2,1)	-.2896272	.4152479	-.697	.486	-1.1034981	.5242438
	UN (2,2)	.0325245	.0373225	.871	.384	.0034312	.3083048

From the estimates of covariance parameters shown in Figure 10-37, you see that, based on the Wald test, you cannot reject the null hypothesis that the variance of the between-subjects slopes is 0, suggesting that a random model with only an intercept term may be adequate. You can obtain this model by removing *age* from the previous random specification or by selecting Compound Symmetry for the Repeated Covariance Type and not specifying a random model. For either specification, the estimates of fixed effects are shown in Figure 10-38. The equations for predicting the distance are:

Distance girls = 17.37 + 0.48 age

Distance boys = 16.34 + 0.78 age

Based on the gender-by-age parameter estimate shown in Figure 10-38, you can reject the null hypothesis that boys and girls have the same slope. Based on the parameter estimate for gender, you cannot reject the null hypothesis that the intercepts are different for boys and girls. Based on the observed significance level for age, you reject the null hypothesis that there is no linear relationship between age and distance.

Figure 10-38
Estimates of fixed effects

Parameter	Estimate	Std. Error	df	t	Sig.	95% Confidence Interval	
						Lower Bound	Upper Bound
Intercept	16.34063	.9813122	103.986	16.652	.000	14.3946430	18.2866070
[GENDER=f]	1.0321023	1.5374208	103.986	.671	.504	-2.0166656	4.0808701
[GENDER=m]	0[1]	0	.	.	.	.	.
AGE	.7843750	.0775011	79	10.121	.000	.6301129	.9386371
[GENDER=f] * AGE	-.3048295	.1214209	79	-2.511	.014	-.5465118	-.0631473
[GENDER=m] * AGE	0[1]	0	.	.	.	.	.

[1]. This parameter is set to zero because it is redundant.

It is important to realize that the hypotheses you are testing in Figure 10-38 and in Table 10-4 are different. In Figure 10-38, you found a significant gender effect, while in Table 10-4, you found the parameter estimate for gender not to be significant. That is not a contradiction. In Table 10-4, you are testing hypotheses about mean distances for boys and girls. You are asking whether the average distances are different for boys and girls. You conclude that they are. In Figure 10-38, you found that boys and girls have significantly different slopes for the regression line that relates distance and age. This tells you that the average values differ for the two groups. The gender test in Figure 10-38 tells you whether the intercepts for the two lines differ.

Advantages of Mixed Models for Analyzing Repeated Measures Designs

Repeated measures designs are often analyzed using multivariate test statistics or pooled univariate tests that are available in the GLM Repeated Measures procedure. In the GLM Repeated Measures procedure, tests are based on constructing a set of linear transformations of the data, such as differences between adjacent time points. All subjects must have measurements at the same time points, and all information must be complete. Subjects with incomplete information are dropped from the analysis. Also, all interactions between within-subject and between-subjects factors must be included in the model. The set of transformations is then analyzed using either multivariate or univariate tests. The univariate within-subjects tests for repeated measures data require that the within-subject variance-covariance matrix be of a specified form—Type H (Huynh and Feldt, 1970). The sphericity test is used to test this assumption. If the sphericity assumption is not violated, univariate tests are more powerful than multivariate tests. If the sphericity assumption is violated, you can adjust the univariate tests with the Greenhouse and Geisser or Huynh and Feldt adjustment. Or you can use multivariate tests, such as Wilks' lambda, Pillai's trace, Hotelling-Lawley Trace, or Roy's Greatest Root. These tests do not make any assumptions about the within-subjects covariance matrix.

Mixed model analysis of repeated measures has several advantages. The time points at which measurements are obtained need not be constant for all subjects. Subjects with incomplete observations can be included in the analysis. These are important considerations in many studies, since people often miss regularly scheduled clinic appointments or experimental sessions. In a mixed model analysis, you can also select the best variance-covariance structure for your data and specify the interactions of interest.

Nonlinear Regression

Many real-world relationships are approximated with linear models, especially in the absence of theoretical models that can serve as guides. You would be unwise to model the relationship between speed of a vehicle and stopping time with a linear model, since the laws of physics dictate otherwise. However, nothing deters you from modeling salary as a linear function of variables such as age, education, and experience. In general, you want to choose the simplest model that fits the observed relationship. Another advantage of linear models is the simplicity of the accompanying statistical estimation and hypothesis testing. Algorithms for estimating parameters of fixed effects linear models under standard assumptions are straightforward with direct solutions that don't require iteration. There are, however, situations in which it is necessary to fit nonlinear models. Before considering the steps involved in nonlinear model estimation, consider what makes a model nonlinear.

What Is a Nonlinear Model?

There is often confusion about the characteristics of a nonlinear model. Consider the following equation:

$$Y = B_0 + B_1 X^2$$

Is this a linear or nonlinear model? The equation is certainly not that of a straight line—it is the equation for a parabola. However, the word *linear*, in this context, does not refer to whether the equation is that of a straight line or a curve. It refers to the functional form of the equation. That is, can the dependent variable be expressed as a linear combination of parameter values and values of the independent variables? The parameters must be linear. The independent variables can be transformed in any fashion. They can be raised to various powers, logged, and so on. The transformation cannot involve the parameters in any way, however.

The previous model is a linear model, since it is nonlinear in only the independent variable X. It is linear in the parameters B_0 and B_1. In fact, you can write the model as

$$Y = B_0 + B_1 X'$$

where X' is the square of X. The parameters in this model can be estimated using the usual linear model techniques.

Transforming Nonlinear Models

Consider the model:

$$Y = e^{B_0 + B_1 X_1 + B_2 X_2 + E}$$

The model, as it stands, is not of the form

$$Y = B_0 + B_1 Z_1 + B_2 Z_2 + \ldots + B_p Z_p + E$$

where the B's are the parameters and the Z's are functions of the independent variables, so it is a nonlinear model. However, if you take natural logs of both sides of the model equation, you get the model

$$\log(Y) = B_0 + B_1 X_1 + B_2 X_2 + E$$

The transformed equation is linear in the parameters, and you can use the usual techniques for estimating them. Models that initially appear to be nonlinear but that can be transformed to a linear form are sometimes called **intrinsically linear models**. It is a good idea to examine what appears to be a nonlinear model to see if it can be transformed to a linear one. Transformation to linearity makes estimation much easier. However, you don't get exactly the same parameter values as when you estimate the parameters from the nonlinear model.

Another example of a transformable nonlinear model is

$$Y = e^B X + E$$

The transformation $B' = e^B$ results in the model

$$Y = B'X + E$$

You can use the usual methods to estimate B' and then take its natural log to get the values of B.

Error Terms in Transformed Models

In both linear and nonlinear models, you assume that the error term is additive. When you transform a model to linearity, you must make sure that the transformed error term satisfies the requisite assumptions. For example, if your original model is

$$Y = e^{BX} + E$$

taking natural logs does not result in a model that has an additive error term. To have an additive error term in the transformed model, your original model would have to be

$$Y = e^{BX + E} = e^{BX}e^E$$

Intrinsically Nonlinear Models

A model such as

$$Y = B_0 + e^{B_1 X_1} + e^{B_2 X_2} + e^{B_3 X_3} + E$$

is **intrinsically nonlinear**. You can't apply a transformation to linearize it. You must estimate the parameters using nonlinear regression. In nonlinear regression, just as in linear regression, you choose values for the parameters so that the sum of squared residuals, or some other function of residuals, is a minimum. There is not, however, a closed solution. You must solve for the values iteratively. There are several algorithms for the estimation of nonlinear models (see Draper and Smith, 1998).

Fitting a Logistic Population Growth Model

For an example of fitting a nonlinear equation, consider a model for population growth. Population growth is often modeled using a logistic population growth model of the form

$$Y_i = \frac{C}{1 + e^{A + BT_i}} + E_i$$

where Y_i is the population size at time T_i. Although the model often fits the observed data reasonably well, the assumptions of independent error and constant variance may be violated because, with time series data, errors are not independent and the size of the error may be dependent on the magnitude of the population. Since the logistic population growth model is not transformable to a linear model, you must use nonlinear regression to estimate the parameters.

Estimating a Nonlinear Model

In nonlinear regression, you must keep changing the values of the parameter estimates until the sum of squared residuals (or other loss function) reaches a minimum. There are different algorithms for deciding which way to turn in the search for the smallest value of the loss function. SPSS has implemented two: the Levenberg-Marquardt method, as proposed by Moré (1977), and the sequential quadratic programming method (Gill et al., 1986).

Examining the Plot

Figure 11-1 is a plot of the decennial population (in millions) of the United States from 1790 to 2000. For the nonlinear regression, the variable *decade*, which is the number of decades since 1790, is the independent variable, instead of the actual year. This is done to prevent possible computational difficulties arising from large data values (see "Computational Issues" on p. 260).

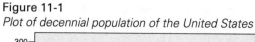

Figure 11-1
Plot of decennial population of the United States

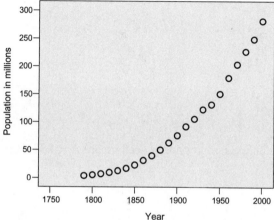

Finding Starting Values

In order to start the nonlinear estimation algorithm, you must have initial values for the parameters. Unfortunately, the results of nonlinear estimation often depend on having good starting values for the parameters. There are several ways of obtaining starting values (see "Estimating Starting Values" on p. 258).

For this example, you can obtain starting values by making some simple assumptions. In the logistic growth model, the parameter C represents the asymptote. You can arbitrarily choose an asymptote that is not too far from the largest observed value. The largest observed value for the population is 281, so you'll start with a value of 300.

Using the value of 300 for C, you can estimate a value for A based on the observed population of 3.93 million at time 0 (1790):

$$3.93 = \frac{300}{1 + e^A}$$

Thus,

$$A = \log\left(\frac{300}{3.93} - 1\right) = 4.32$$

To estimate a value for B, you can use the population at time 1 (5.31 million) and the estimates of C and A. This gives

$$5.31 = \frac{300}{1 + e^{B + 4.32}}$$

from which you derive

$$B = \log\left(\frac{300}{5.31} - 1\right) - 4.32 = -0.30$$

The initial values in the nonlinear regression procedure are A = 4.32, B = –0.30, and C = 300.

Specifying the Analysis

To obtain a nonlinear regression analysis in SPSS, you must:

■ Assign names and starting values to each parameter.

■ Enter the formula for the nonlinear equation, using the assigned parameter names.

Open the data file *census.sav*, and from the menus choose:

Analyze
  Regression
    Nonlinear...

Then make the dialog box selections as shown in Figure 11-2.

Figure 11-2
Nonlinear Regression dialog box

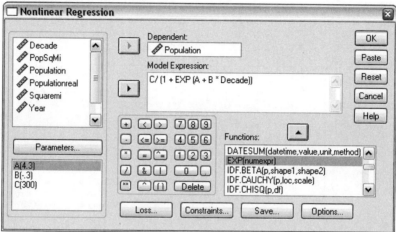

Parameters are assigned names and starting values in the Parameters dialog box. To save residuals and predicted values, in the Save dialog box select Predicted values and Residuals.

Estimating the Parameters

Figure 11-3 shows the residual sums of squares and parameter estimates at each iteration. At step 1, the parameter estimates are the starting values you set. At the major iterations, which are identified with integer numbers, the derivatives are evaluated and the direction of the search is determined. At the minor iterations, the distance is established. As the note at the end of the table indicates, iteration stops when the relative change in residual sums of squares between iterations is less than or equal to the convergence criterion or when the number of specified iterations is reached. Always make sure that the algorithm converged to a solution instead of stopping at the specified maximum number of iterations.

Figure 11-3

Parameter estimates

Iteration History[2]

Iteration Number[1]	Residual Sum of Squares	Parameter		
		A	B	C
1.0	4854.257	4.300	-.300	300.000
1.1	5625.440	3.679	-.206	355.212
1.2	1479.740	4.115	-.254	327.143
2.0	1479.740	4.115	-.254	327.143
2.1	1937.480	3.904	-.210	409.380
2.2	738.485	4.055	-.233	378.370
3.0	738.485	4.055	-.233	378.370
3.1	593.989	3.997	-.213	434.331
4.0	593.989	3.997	-.213	434.331
4.1	455.601	4.035	-.215	444.155
5.0	455.601	4.035	-.215	444.155
5.1	455.554	4.033	-.215	445.550
6.0	455.554	4.033	-.215	445.550
6.1	455.553	4.034	-.215	445.481
7.0	455.553	4.034	-.215	445.481
7.1	455.553	4.034	-.215	445.491

Derivatives are calculated numerically.

[1]. Major iteration number is displayed to the left of the decimal, and minor iteration number is to the right of the decimal.

[2]. Run stopped after 16 model evaluations and 7 derivative evaluations because the relative reduction between successive residual sums of squares is at most SSCON = 1.00E-008.

Summary statistics for the nonlinear regression are shown in Figure 11-4. For a nonlinear model, the tests used for linear models are not appropriate. In this situation, the residual mean square is not an unbiased estimate of the error variance, even if the model is correct. For practical purposes, you can still compare the residual variance with an estimate of the total variance, but the usual F statistic cannot be used for testing hypotheses.

The entry in Figure 11-4 labeled *Uncorrected Total* is the sum of the squared values of the dependent variable. The entry labeled *(Corrected Total)* is the sum of squared deviations around the mean. The *Regression* sum of squares is the sum of the squared predicted values.

Figure 11-4

Summary statistics for model

ANOVA[1]

Source	Sum of Squares	df	Mean Squares
Regression	356175.77	3	118725.258
Residual	455.553	19	23.976
Uncorrected Total	356631.33	22	
Corrected Total	159649.41	21	

Dependent variable: Population in millions

[1]. R squared = 1 - (Residual Sum of Squares) / (Corrected Sum of Squares) = .997.

The entry labeled *R squared* is the coefficient of determination. It may be interpreted as the proportion of the total variation of the dependent variable around its mean that is explained by the fitted model. For nonlinear models, its value can be negative if the selected model fits worse than the mean. Figure 11-5 is a plot of the observed and predicted populations.

Figure 11-5

Plot of observed and predicted population values

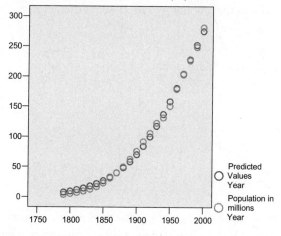

Approximate Confidence Intervals for the Parameters

In the case of nonlinear regression, it is not possible to obtain exact confidence intervals for each of the parameters. Instead, you must rely on **asymptotic** (large sample) approximations or on bootstrapped estimates. Figure 11-6 shows the estimated parameters, asymptotic standard errors, and asymptotic 95% confidence intervals.

Figure 11-6
Estimated parameters and asymptotic confidence intervals

Parameter	Estimate	Std. Error	95% Confidence Interval	
			Lower Bound	Upper Bound
A	4.034	.067	3.892	4.175
B	-.215	.010	-.236	-.194
C	445.491	36.036	370.067	520.916

The asymptotic correlation matrix of the parameter estimates is shown in Figure 11-7. If there are very large positive or negative values for the correlation coefficients, it is possible that the model is **overparameterized**. That is, a model with fewer parameters may fit the observed data as well. This does not necessarily mean that the model is inappropriate; it may mean that the amount of data is not sufficient to estimate all of the parameters.

Figure 11-7
Asymptotic correlation matrix of parameter estimates

	A	B	C
A	1.000	-.494	-.116
B	-.494	1.000	.916
C	-.116	.916	1.000

Bootstrapped Estimates

You can also obtain bootstrapped estimates of the standard errors of the parameters as well as confidence intervals. Bootstrapped estimates are calculated by taking repeated samples with replacement from the original data set. The number of cases is the same as in the original data set; the number of samples used in the nonlinear regression procedure is $10p(p + 1) / 2$, where p is the number of parameters. The nonlinear equation is estimated for each of these samples. The standard error of each parameter estimate is then calculated as the standard deviation of the bootstrapped estimates of that parameter. Parameter values from the original data are used as starting values for each bootstrap sample. The sequential quadratic programming algorithm is used if you select bootstrapped estimates.

To obtain bootstrapped estimates, recall the dialog box and select:

Options...
   ☑ Bootstrapped estimates of standard error

The parameter estimates and their bootstrapped standard errors are shown in Figure 11-8.

Figure 11-8
Parameters and bootstrapped estimates

	Parameter	Estimate	Std. Error	95% Confidence Interval		95% Trimmed Range	
				Lower Bound	Upper Bound	Lower Bound	Upper Bound
Bootstrap[1,2]	A	4.034	.067	3.900	4.167	3.919	4.151
	B	-.215	.012	-.239	-.191	-.249	-.197
	C	445.491	47.185	351.073	539.908	342.584	513.899

[1.] Based on 60 samples.

[2.] Loss function value equals 455.553.

Examining the Residuals

The SPSS Nonlinear Regression procedure allows you to save predicted values and residuals that can be used for exploring the goodness of fit of the model. Figure 11-9 is a plot of residuals against the observed year values. You will note that the errors appear to be correlated and that the variance of the residuals increases with time.

Figure 11-9
Plot of residuals against time

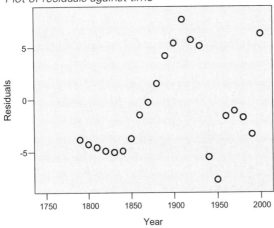

Estimating Starting Values

Good initial values are important in nonlinear regression and may provide a better solution in fewer iterations. In addition, computational difficulties can sometimes be avoided by a good choice of initial values. Poor initial values can result in nonconvergence, a local rather than global solution, or a physically impossible solution. If you don't have starting values, don't just set them all to 0. Use values in the neighborhood of what you expect to see. There are a number of ways to determine initial values for nonlinear models. A combination of techniques can be used.

Use Starting Values from Previous Analysis

If you have already run a nonlinear regression from the dialog box, you can select this option in the Parameters dialog box to obtain the initial values of parameters from their values in the previous run. This permits you to continue searching when the algorithm is converging slowly. (The initial starting values will still appear on the Parameters list in the main dialog box.) This selection persists in this dialog box for the rest of your session. *If you change the model, be sure to deselect it.*

Look for a Linear Approximation

If you ignore the error term, sometimes a linear form of the model can be derived and then linear regression can be used to obtain initial values. For example, consider the model

$$Y = e^{A+BX} + E$$

If you ignore the error term and take the natural log of both sides of the equation, you obtain the model

$$\log(Y) = A + BX$$

You can use linear regression to estimate *A* and *B* and specify these values as starting values in nonlinear regression.

Use Properties of the Nonlinear Model

Sometimes you know the values of the dependent variable for certain combinations of parameter values. For example, if in the model

$$Y = e^{A+BX}$$

you know that when *X* is 0, *Y* is 2, you would select the natural log of 2 as a starting value for *A*. Examination of an equation at its maximum, minimum, and when all the independent variables approach 0 or infinity may help in selection of initial values.

Solve a System of Equations

By taking as many data points as you have parameters, you can solve a simultaneous system of equations. For example, in the previous model, you can solve the equations

$$\log(Y_1) = A + BX_1$$
$$\log(Y_2) = A + BX_2$$

Using subtraction,

$$\log(Y_1) - \log(Y_2) = BX_1 - BX_2$$

You can solve for the values of the parameters:

$$B = \frac{\log(Y_1) - \log(Y_2)}{X_1 - X_2}$$

and

$$A = \log(Y_1) - BX_1$$

Computational Issues

Computationally, nonlinear regression problems can be difficult to solve. Models that require exponentiation or powers of large data values may cause underflows or overflows. (An **overflow** is caused by a number that is too large for the computer to handle, while an **underflow** is caused by a number that is too small for the computer to handle.) Sometimes the program may continue and produce a reasonable solution, especially if only a few data points caused the problem. If this is not the case, you must eliminate the cause of the problem. If your data values are large—for example, years—you can subtract the smallest year from all of the values. That's what was done with the population example. Instead of using the actual years, we used decades since 1790. (To compute the number of decades, you subtracted the smallest year from each year value and divided the result by 10.) You must, however, consider the effect of rescaling on the parameter values. Many nonlinear models are not scale invariant. You can also consider rescaling the parameter values.

If the program fails to arrive at a solution—that is, if it doesn't converge—you might consider choosing different starting values. You can also change the criterion used for convergence.

If none of these strategies works, you can try a different algorithm. By default, the Levenberg-Marquardt algorithm is used. The algorithm combines linear descent at early iterations and then uses the Gauss-Newton method for subsequent iterations. You can choose the sequential quadratic programming algorithm for problems that are causing difficulties for the default algorithm. For a particular problem, one algorithm may perform better than the other.

Additional Nonlinear Regression Options

Additional options are available for nonlinear models when the sequential quadratic programming algorithm is used:

■ You can supply linear and nonlinear constraints for the values of the parameter estimates. For example, you can specify that a parameter value must be greater than 0 or that it must fall in the interval from 0 to 2.

■ You can specify your own loss function. The loss function is the function that is minimized by the algorithm. The default loss function is the sum of the squared residuals. That means you want to find parameter estimates that make the sum of squared residuals as small as possible. Most loss functions involve residuals, and the variable *RESID_* is available in the variable list.

Nonlinear Regression Common Models

The table below provides example model syntax for many published nonlinear regression models. *A model selected at random is not likely to fit your data well.* Appropriate starting values for the parameters are necessary, and some models require constraints in order to converge.

Table 11-1

Example model syntax

Name	Model expression
Asymptotic Regression	$b1 + b2 * \exp(b3 * x)$
Asymptotic Regression	$b1 - (b2 * (b3 ** x))$
Density	$(b1 + b2 * x) ** (-1 / b3)$
Gauss	$b1 * (1 - b3 * \exp(-b2 * x ** 2))$
Gompertz	$b1 * \exp(-b2 * \exp(-b3 * x))$
Johnson-Schumacher	$b1 * \exp(-b2 / (x + b3))$
Log-Modified	$(b1 + b3 * x) ** b2$
Log-Logistic	$b1 - \ln(1 + b2 * \exp(-b3 * x))$
Metcherlich Law of Diminishing Returns	$b1 + b2 * \exp(-b3 * x)$
Michaelis Menten	$b1 * x / (x + b2)$
Morgan-Mercer-Florin	$(b1 * b2 + b3 * x ** b4) / (b2 + x ** b4)$
Peal-Reed	$b1 / (1 + b2 * \exp(-(b3 * x + b4 * x ** 2 + b5 * x ** 3)))$
Ratio of Cubics	$(b1 + b2 * x + b3 * x ** 2 + b4 * x ** 3) / (b5 * x ** 3)$

Table 11-1
Example model syntax

Ratio of Quadratics	(b1 + b2 * x + b3 * x ** 2) / (b4 * x ** 2)
Richards	b1 / ((1 + b3 * exp(– b2 * x)) ** (1 / b4))
Verhulst	b1 / (1 + b3 * exp(– b2 * x))
Von Bertalanffy	(b1 ** (1 – b4) – b2 * exp(–b3 * x)) ** (1 / (1 – b4))
Weibull	b1 – b2 * exp(– b3 * x ** b4)
Yield Density	(b1 + b2 * x + b3 * x ** 2) ** (–1)

Specifying a Segmented Model

You can also define a model that changes for different parts of the data. For example, you can fit the models:

$$Y_i = \frac{C}{1 + e^{A + BT_i}} \quad \text{if } 0 < t < 15$$

$$Y_i = t \qquad\qquad \text{if } t > = 15$$

To specify such a segmented model, you must write the model as a sum of a series of terms, one for each condition. Each term consists of a logical expression (in parentheses) multiplied by the expression that should result when that logical expression is true.

For example, consider the following segmented model:

$$Y_i = 0 \quad \text{if } X < = 0$$

$$Y_i = X \quad \text{if } 0 < X < 1$$

$$Y_i = 1 \quad \text{if } X > = 1$$

The model can be written as

$$(X < = 0) * 0 + (X > 0 \ \& \ X < 1) * X + (X > = 1) * 1$$

The logical expressions in parentheses have values of 1 if the expression is true or 0 if the expression is false. For example, if $X < = 0$, the above reduces to
$1*0 + 0*X + 0*1 = 0$.

If $0 < X < 1$, it reduces to $0*0 + 1*X + 0*1 = X$

If $X >= 1$, it reduces to $0*0 + 0*X + 1*1 = 1$

More complicated examples can be easily built by substituting different logical expressions and outcome expressions. Remember that double inequalities, such as $0 < X < 1$, must be written as compound expressions, such as $(X > 0 \& X < 1)$.

String variables can be used within logical expressions:

(city = 'New York')*costliv + (city = 'Des Moines')*0.59*costliv

This yields one expression (the value of the variable *costliv*) for New Yorkers and another (59% of that value) for Des Moines residents. String constants must be enclosed in quotation marks or apostrophes, as shown here.

Two-Stage
Least-Squares Regression

In ordinary least-squares regression, you must assume that errors are independent of the predictor variables. This need not be the case. For example, macroeconomic data—data describing the overall state of the economy—frequently take the form of time series. Because of the complex interrelationships among macroeconomic variables, models for such data are usually afflicted with correlated errors—the errors in the equation are correlated with one or more of the predictor variables.

When predictor variables are correlated with errors, results obtained from ordinary least-squares (OLS) regression are biased. In this chapter, you will use the technique known as two-stage least squares to deal with correlated errors using a classic macroeconomic model in a very modest setting.

Artichoke Data

To illustrate the concepts involved in two-stage least squares, consider a hypothetical data set from Kelejian and Oates (1989) for the production of artichokes. Three main series are involved in the model: the demand for artichokes, expressed as the quantity sold in tons; the price of an artichoke in cents; and the average family income in thousands of dollars. Figure 12-1 is a sequence chart of the three variables (Graph menu, Sequence Charts).

Figure 12-1
Demand, price, and income

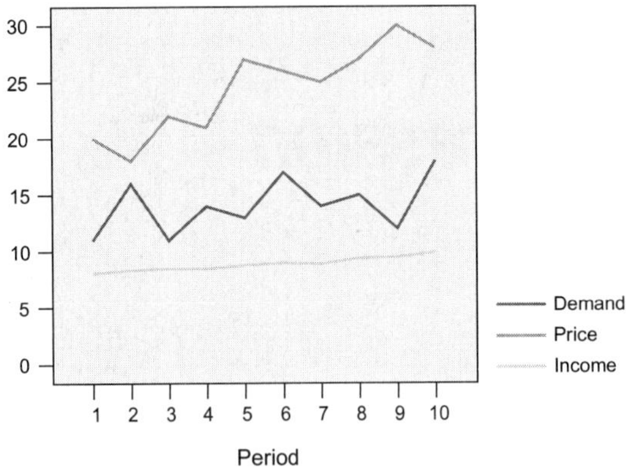

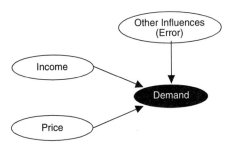

Demand-Price-Income Economic Model

A classic economic model of the demand for some commodity, such as artichokes, is

$$\text{demand} = \beta_0 + \beta_1 \times \text{price} + \beta_2 \times \text{income} + \text{error}$$

The path diagram in Figure 12-2 represents this equation.

Figure 12-2
Path diagram for simple model

Although there is more to the model than this one equation, let's go ahead and estimate the coefficients using ordinary least-squares regression.

Estimation with Ordinary Least Squares

Figure 12-3 shows ordinary least-squares (OLS) estimation of the previously described model. To obtain the results, open the data file *artichoke.sav* and from the menus choose:

Analyze
  Regression
    Linear...

▶ Dependent: demand
▶ Independent(s): price income

This model estimates that an extra thousand dollars of mean family income increases demand for artichokes by about 6.2 tons, while each extra penny in the price of artichokes decreases demand by about 0.66 tons. Since Kelejian and Oates' data are hypothetical, we will not dwell on the interpretation of the coefficients. Instead, let us consider some difficulties with such models.

Figure 12-3
Ordinary least-squares results

Model	R	R Square	Adjusted R Square	Std. Error of the Estimate
1	.738[1]	.545	.415	1.854

[1]. Predictors: (Constant), income, price

Model		Unstandardized Coefficients		Standardized Coefficients		
		B	Std. Error	Beta	t	Sig.
1	(Constant)	-25.082	13.551		-1.851	.107
	price	-.659	.317	-1.066	-2.079	.076
	income	6.209	2.212	1.439	2.807	.026

Feedback and Correlated Errors

The difficulty with a model relating production, price, and income is that the influences work in both directions. The model states that the quantity of artichokes produced depends upon the price—when prices are high, farmers tend to grow more artichokes. It is equally true that the price depends upon production—a glut of artichokes on the market will force prices back down. (The model assumes that, at the market price, the demand for artichokes equals supply.) Thus, we should regard the previous model as part of a system of interrelated equations. The OLS solution shown above ignored the feedback effect of production on price. The real situation is more like the one shown in Figure 12-4.

Figure 12-4
Path diagram showing feedback

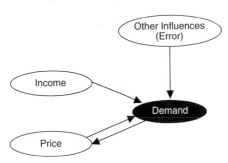

Suppose that something not included in the model (a new fertilizer, perhaps) leads to an increased quantity of artichokes on the market. Prices will fall due to the increased production. Because of the feedback relationship, high values of the error term in Figure 12-4 (which represents the effect of things not included in the equation) would be associated with low prices of artichokes.

Correlation of the error term with one of the predictor variables violates one of the assumptions of regression analysis. It leads to biased coefficients because the model in Figure 12-2 implies that price levels cause increased production. The OLS algorithm used by the Linear Regression procedure treats that portion of the error that is correlated with *price* as being caused by *price*—although really the correlation arises in the other direction, from the feedback effect of *demand* on *price*.

Note that it is the theoretical errors that are correlated with the predictor *price*. If you actually use the Linear Regression procedure, the OLS algorithm assumes the errors to be uncorrelated with the predictors and calculates biased coefficients in such a way as to produce uncorrelated residuals. Thus, you cannot use the estimated errors produced by a regression procedure to check for the problem of correlated errors. You know the correlated errors are there because of the feedback loop in the correctly specified model.

Two-Stage Least Squares

You know there are problems with the ordinary least-squares analysis presented above. The feedback relationship from the dependent variable *demand* to the predictor variable *price* produces correlations between the error term and *price*, and therefore the estimates from OLS regression are biased.

Two-stage least squares (2SLS) is an important regression technique for models in which one (or more) of the predictor variables is thought to be correlated with the error term. Before we discuss 2SLS strategy, we need to introduce some terminology.

Endogenous variables. Endogenous literally means *produced from within*. In regression analysis, an endogenous variable is a variable that is causally dependent on the other variable(s) in the model. When you are specifying models that solve several equations simultaneously (whether or not you explicitly solve all the equations), you know that several endogenous variables are present.

In a feedback situation, each of the variables in the feedback relationship is endogenous. Thus, in the model, *price* and *demand* are endogenous variables. The model should include the two-way relationship between *price* and *demand*, as in Figure 12-4.

Instruments. Instrumental variables, or simply *instruments,* are variables that are not influenced by other variables in the model but that do influence those variables. They may or may not be a part of the equation you are interested in, but they must be free of causal influence from any of the variables in that equation. To be effective, instruments should be:

- Highly correlated with the endogenous variables
- Not correlated with the error terms

Another practical consideration is that the instrumental variables must be available for use in your analysis.

In Figure 12-5, you can see that there is a path *from* the instruments to the endogenous variable *price*, but there are no paths leading *to* the instruments from the rest of the model.

Figure 12-5
Instrument variables

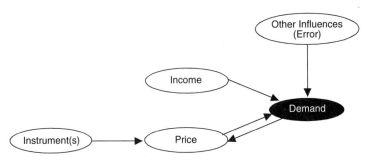

In practice, it is difficult to be sure whether an instrument is correlated with the (unobserved) error term. As noted in "Feedback and Correlated Errors" on p. 267, you cannot test this by using the estimated error terms from the Linear Regression procedure. When no instrument is readily available, the lagged value of the endogenous variable is often used. Even if this is correlated with the lagged error, it may not be correlated with the current error.

Strategy

In two-stage least squares, an endogenous predictor variable (*price*) is correlated with the theoretical error terms in your model for the dependent variable (*demand*). OLS wrongly attributes some of the theoretically unexplained variation in *demand* to the effect of *price* because of this correlation. The 2SLS strategy is to replace the troublesome endogenous predictor variable *price* with a similar variable that:

■ Is almost as good as *price* at predicting *demand*.

■ Is not correlated with the theoretical error term in the prediction of *demand*.

You obtain such a replacement variable by ordinary regression, using the instruments to predict the endogenous variable. If the instruments have the two properties listed in "Two-Stage Least Squares" on p. 269, the predicted value of the endogenous variable will be:

■ A good predictor of the dependent variable.

■ Uncorrelated with the error term for the dependent variable.

To appreciate what is involved in two-stage least squares, you will work through the very simple demand-for-artichokes example using the Linear Regression procedure (in two stages). You will then see how the 2-Stage Least Squares procedure automates the process.

Stage 1: Estimating Price

The first stage requires instruments with which to predict *price*. If the instruments are not affected by *demand*, then the predicted values of *price* will likewise be unaffected by *demand*, and we can safely use those predicted values in the second stage.

The instruments you will use are:

- *income.* This is one of the predictor variables in the original model. It is unlikely that demand for artichokes affects the income of consumers in any large way, but it is possible that income levels will be useful in predicting price levels.

- *rainfall.* This variable is not a part of the original model because we do not expect it to be useful in predicting demand for artichokes. It should affect the price of artichokes, however, because of its effect on the quantity available. Since it is unlikely to be influenced by *demand* (or anything else), it is a suitable instrument.

- Lagged *price.* The lagged value of an endogenous variable is very often used as an instrument because it is frequently a good predictor of the current value. You have good reason to believe that current demand for artichokes does not affect last season's price, so lagged *price* should be uncorrelated with the errors in the demand model. (To create *lagprice*, from the Transform menu choose Compute and enter lagprice=lag(price).)

The first stage of the estimation, using these instruments, is shown in Figure 12-6. The predicted values of *price* from this equation are saved in a new series named *pre_1*. To obtain the output, recall the Linear Regression dialog box and select:

▶ Dependent: price
▶ Independent(s): income, rainfall, lagprice

Save...
   Predicted Values
   ☑ Unstandardized

Figure 12-6
Using instruments to predict price

Model	R	R Square	Adjusted R Square	Std. Error of the Estimate
1	.919[1]	.844	.750	1.911

[1]. Predictors: (Constant), lagprice, rainfall, income

Model		Unstandardized Coefficients		Standardized Coefficients	t	Sig.
		B	Std. Error	Beta		
1	(Constant)	-8.580	20.892		-.411	.698
	income	3.752	2.994	.506	1.253	.266
	rainfall	-.218	.106	-.442	-2.062	.094
	lagprice	.418	.418	.431	1.001	.363

Stage 2: Estimating the Model

The second stage is to estimate the model with *pre_1* (the predicted value for *price* from the previous model) substituting for *price*. The regression coefficients in Figure 12-7 are the ones you want, given what you know about feedback in the overall model. The coefficient of *income* is 9.561, and the coefficient of *price* is –1.265. Notice that both coefficients have changed appreciably from Figure 12-3, where you used ordinary least squares to estimate them.

Figure 12-7
Using predicted price to predict demand

Model	R	R Square	Adjusted R Square	Std. Error of the Estimate
1	.857[1]	.734	.645	1.369

[1]. Predictors: (Constant), pre_1, income

Model		Unstandardized Coefficients		Standardized Coefficients	t	Sig.
		B	Std. Error	Beta		
1	(Constant)	-40.017	13.708		-2.919	.027
	income	9.561	2.354	2.147	4.062	.007
	pre_1	-1.265	.346	-1.933	-3.658	.011

However, the R^2 and the standard errors reported in Figure 12-7 are not the ones you want. This is because Figure 12-7 shows an equation with *pre_1*, but you want the R^2

and standard errors that use *price*. The regression coefficients for the two equations are the same, but the R^2 and the standard errors are not. In order to obtain the correct standard errors, you must run the 2-Stage Least Squares procedure.

2-Stage Least Squares Procedure

To obtain the 2-Stage Least Squares solution directly, from the menus choose:

Analyze
  Regression
    2-Stage Least Squares...

▶ Dependent: demand
▶ Explanatory: price, income
▶ Instrumental: rainfall, lagprice, income

Figure 12-8 and Figure 12-9 show the 2-Stage Least Squares procedure results, without going through the two stages outlined above.

Figure 12-8
2-Stage Least Squares solution

Equation 1	Multiple R	.681
	R Square	.464
	Adjusted R Square	.285
	Std. Error of the Estimate	2.442

Figure 12-9
2-Stage Least Squares solution

		Unstandardized Coefficients		Beta	t	Sig.
		B	Std. Error			
Equation 1	(Constant)	-40.017	24.452		-1.637	.153
	price	-1.265	.617	-2.105	-2.051	.086
	income	9.561	4.199	2.147	2.277	.063

The 2-Stage Least Squares procedure results in the same coefficient estimates as two stages of OLS, with much less effort. The contrast in effort required is even more striking in larger models.

The 2-Stage Least Squares procedure is quite simple to specify. You must, however, understand the model you are estimating if you are to specify it correctly.

13

Weighted Least-Squares Regression

When you estimate the parameters of a linear regression model, all observations usually contribute equally to the computations. This is called **ordinary least-squares (OLS) regression**. When all of the observations have the same variance, this is the best strategy, since it results in parameter estimates that have the smallest possible variances. However, if the observations are not measured with equal precision, OLS no longer yields parameter estimates with the smallest variance. A modification known as **weighted least-squares (WLS) regression** does. In weighted least-squares regression, data points are weighted by the reciprocal of their population variances. This means that observations with large variances have less impact on the analysis than observations associated with small variances.

Diagnosing the Problem

As an example of the use of weighted least-squares, consider data collected by W. M. Hinshaw and presented in Draper and Smith (1998). As always, the first step is to plot the values of the two variables, as shown in Figure 13-1. One of the first things you notice is that the variability (or spread) of the dependent variable increases with increasing values of the independent variable. This indicates that the assumption of equal variances across all data points is probably violated and the ordinary least-squares approach is no longer optimal.

Figure 13-1
Scatterplot of y and x

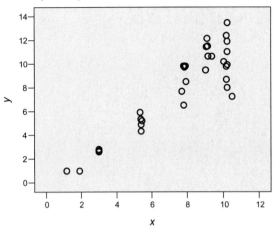

If you fit an ordinary least-squares regression to the data points (by running the Linear Regression procedure), you obtain the output shown in Figure 13-2.

Figure 13-2
Linear Regression coefficients

Model		Unstandardized Coefficients		Standardized Coefficients	t	Sig.
		B	Std. Error	Beta		
1	(Constant)	-.579	.679		-.852	.400
	x	1.135	.086	.917	13.169	.000

You can't tell from the coefficients that the requisite regression assumptions are violated. You must examine the residual plots. Figure 13-3, from the Linear Regression procedure, is a plot of the Studentized residuals against the predicted values of the dependent variable.

Figure 13-3
Plot of residuals and predicted values

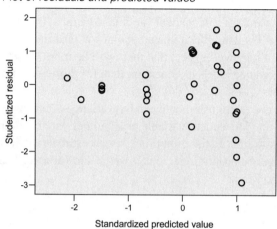

You see that instead of being randomly distributed around the line through 0, the residuals form a funnel, suggesting again that the variances are unequal and that the data must be transformed to achieve equal variances or that weighted least-squares should be used.

Estimating the Weights

Estimation of the regression model with weighted least-squares requires an estimate of the relative variability at each point. That is, although you don't need to know the actual variance at each point, you must know the ratio of the variances at all points. Unfortunately, this information is often unavailable, and you must estimate the variability from the data. The consequence of estimating the weights from the data is that the weighted least-squares solution no longer results in parameter estimates that necessarily have minimum variance.

You can estimate weights by examining the relationship of the variance to various powers of values of a selected variable. If you have multiple observations (replicates) for each combination of values of the independent variable, you can compute the variance for each set of replicates and use the estimated variance to weight the data points.

Estimating Weights as Powers

It is not unusual for the variance of a dependent variable to be related to the magnitude of an independent variable. For example, if you are looking at the relationship between income and education, you may well expect that there will be more variability in income for people with graduate levels of education than for those who did not complete grammar school.

If you think that there is a relationship between the variance of the dependent variable and the value of an independent variable or any other variable you have available, you can use the Weight Estimation procedure to estimate the weights. However, the variance must be proportional to a power of the variable. That is, the relationship must be of the form:

variance $\propto$ variable^{power}

You can specify a power range and an increment, and the program will evaluate the log-likelihood function for all powers within the grid and then select the power corresponding to the largest log-likelihood.

In the Draper and Smith example, if you group cases with similar values for the independent variable and compute their standard deviations, you obtain the plot shown in Figure 13-4.

Figure 13-4

Plot of standard deviation of y with x

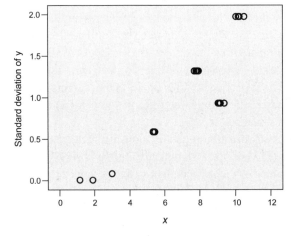

It appears that the standard deviation of *y* is linearly related to *x*. That means the variance is related to the square of *x*. You can evaluate the effect of using different weights by looking at the log-likelihood values that result from different powers. You want to identify a sensible value that corresponds to a maximum for the log-likelihood.

Specifying the Analysis

To run the Weight Estimation procedure with *x* as the weight variable and the power ranging from 0 to 3 in increments of 0.2, from the menus choose:

Analyze
  Regression
    Weight Estimation...

▶ Dependent: y
▶ Independent(s): x
▶ Weight Variable: x
  Power range: 0 through 3 by 0.2

☑ Include constant in equation

Options...
☑ Save best weight as new variable
  Display ANOVA and Estimates
  ⊙ For best power

Examining the Log-Likelihood Functions

The values of the log-likelihood function for various powers of *x* are shown in Figure 13-5. The largest value of the log-likelihood is for a power of 2, confirming the observation that the variance is related to the square of *x*.

Figure 13-5

Log-likelihood functions for powers from WLS

Log-Likelihood Values[2]

Power		
Power	.000	-61.827
	.200	-60.592
	.400	-59.386
	.600	-58.220
	.800	-57.108
	1.000	-56.067
	1.200	-55.119
	1.400	-54.291
	1.600	-53.615
	1.800	-53.131
	2.000	-52.877[1]
	2.200	-52.897
	2.400	-53.228
	2.600	-53.901
	2.800	-54.932
	3.000	-56.322

1. The corresponding power is selected for further analysis because it maximizes the log-likelihood function.

2. Dependent variable: y, source variable: x

WLS Solutions

Figure 13-6 and Figure 13-7 show the WLS solution when the value of 2 is used for the power.

Figure 13-6

Model summary

Multiple R	.974
R Square	.948
Adjusted R Square	.947
Std. Error of the Estimate	.173
Log-likelihood Function Value	-52.877

Figure 13-7

Model coefficients

	Unstandardized Coefficients		Standardized Coefficients			
	B	Std. Error	Beta	Std. Error	t	Sig.
(Constant)	-.580	.190			-3.053	.004
x	1.130	.046	.974	.040	24.632	.000

The weights that are saved from the procedure are the reciprocals of x^2. You will note that, compared to the OLS results in Figure 13-2, the parameter estimates for the slope and intercept have not changed much, but their standard errors have. In the OLS solution, the standard error of the slope (*SE B*) is 0.086. In the WLS solution, the standard error of the slope is 0.046. Similarly, for the constant, the standard error has changed from 0.68 to 0.19.

Estimating Weights from Replicates

If you have replicates in your data—that is, groups of cases for which the values of the independent variable are the same or similar—you can compute the variances of the dependent variable for all of the distinct combinations of the independent variables. The reciprocal of the variances is then the weight, since you want points associated with large variances to have less impact than points with smaller variances. If you have few observations at each point, the variance estimates based on replicates may not be very reliable.

Diagnostics from the Linear Regression Procedure

The Weight Estimation procedure will only estimate weights and provide summary regression statistics. You must use the Linear Regression procedure specifying the weight variable to obtain residuals and other diagnostic information. In fact, if you know the weights, there is no need to run the Weight Estimation procedure; you can perform weighted least-squares analyses using the Linear Regression procedure.

Draper and Smith estimated the weights for the example data by computing variances for cases with similar values of the independent variables. They then developed a quadratic regression model to predict the variances from the values of the independent variable. If you apply their weights using the Linear Regression procedure, you will obtain the summary statistics shown in Figure 13-8.

Figure 13-8

Coefficients from the Draper and Smith estimated weights

Coefficients[1,2]

Model		Unstandardized Coefficients		Standardized Coefficients	t	Sig.
		B	Std. Error	Beta		
1	(Constant)	-.889	.300		-2.963	.006
	x	1.165	.059	.960	19.629	.000

[1.] Dependent Variable: y

[2.] Weighted Least Squares Regression - Weighted by wt

Note that the slope has changed somewhat from the OLS solution, as has the constant.

This solution also differs from that shown in Figure 13-6. One of the reasons for this difference is that Draper and Smith's weights are not as extreme as those estimated by the WLS procedure using a quadratic source model. Our estimated weights range from 0.75 to 0.0076, while Draper and Smith's weights range from 0.30 to 7.8. (The actual numbers used for the weights don't matter. All that matters is the proportionality.) The Draper and Smith weights are probably more realistic, since our estimated hundredfold increase in variance over the observed range of the independent variable is probably too extreme.

To see how well the Draper and Smith WLS solution performs, let's examine the residuals. You must save the residuals and predicted values from the Linear Regression procedure, and you must transform them before plotting. They must be multiplied by the square root of the weight variable as described in the warning in Figure 13-9.

Figure 13-9
Warning from Linear Regression procedure

Warnings

> No plots are produced for Weighted Least Squares regression. You can SAVE the appropriate variables and use other procedures (e.g., EXAMINE and PLOT) to produce the requested plots. To plot weighted versions of the residuals and predicted values, use COMPUTE before plotting: COMPUTE RESID = SQRT(REGWGTvar) * RESID COMPUTE PRED = SQRT(REGWGTvar) * PRED.

Figure 13-10 is a plot of the transformed residuals and predicted values.

Figure 13-10
Plot of transformed residuals and predicted values

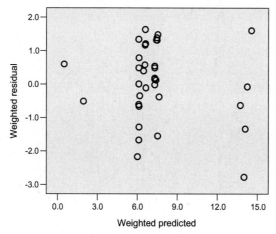

You see that the funnel shape that was evident in Figure 13-3 is no longer as marked. It appears that the WLS solution was successful.

Multidimensional Scaling

This chapter was written by Forrest W. Young and David F. Harris of the Psychometric Laboratory, University of North Carolina.

What characteristics of an automobile do people consider when they are deciding which car to buy: Its economy? Its sportiness? Its reliability? What aspects of a political candidate are important when a voter is making a decision: The candidate's party? The candidate's position on defense issues? What interpersonal characteristics come into play when one member of a work group is talking to another member: The status of the two members? Their knowledge of the task of the work group? Their socioeconomic characteristics?

How do you go about answering such questions? The variables mentioned in the first paragraph are subjective, as are their units of measure and the values of these units. These variables are presumed to exist in the minds of people and do not have an independent, objective existence.

There are at least two ways of answering the questions posed above—at least two ways to construct objective scales that can be reasonably thought to correspond to a person's internal "scales." One of these ways is to obtain multivariate data and then use factor analysis. For more information, see the chapter "Factor Analysis" in the *SPSS 14.0 Statistical Procedures Companion* (Norušis, 2005).

The other way is to obtain dissimilarity data and then use multidimensional scaling to analyze the data. In this chapter, we focus on the multidimensional scaling of dissimilarity data as a way to construct objective scales of subjective attributes.

Data, Models, and Analysis of Multidimensional Scaling

What is multidimensional scaling? **Multidimensional scaling (MDS)** is designed to analyze distance-like data called **dissimilarity data,** or data that indicate the degree of dissimilarity (or similarity) of two things. For example, MDS could be used with data that indicate the apparent (dis)similarity of a pair of political candidates or a pair of automobiles. MDS analyzes the dissimilarity data in a way that displays the structure of the distance-like data as a geometrical picture. MDS has its origins in psychometrics, where it was proposed to help understand people's judgments of the similarity of members of a set of objects. MDS has now become a general data analysis technique used in a wide variety of fields.

What are dissimilarities data? MDS pictures the structure of a set of objects from data that approximate the distances between pairs of the objects. The data, which are called similarities, dissimilarities, distances, or proximities, must reflect the amount of (dis)similarity between pairs of objects. In this chapter, we use the term *dissimilarity* generically to refer to both similarities (where large numbers indicate great similarity) and dissimilarities (where large numbers indicate great dissimilarity).

Traditionally, dissimilarity data are subjective data obtained by asking people to judge the dissimilarity of pairs of things, such as automobiles, political candidates, types of wine, and so on. But dissimilarity data can also be objective measures, such as the driving time between two cities or the frequency with which pairs of people in a work group talk to each other. Dissimilarity data can also be calculated from multivariate data, as when we use voting records in the United States Senate to calculate the proportion of agreement in the votes cast by pairs of senators. However, the data must always represent the degree of dissimilarity of pairs of objects or events.

There may be one or more matrices of dissimilarity data. For example, if we have observed the frequency with which pairs of people in a business communicate with each other, then we have a single similarity matrix where the frequency of communication indicates the similarity between the pair of people. On the other hand, if we have judgments from many drivers of the dissimilarity between pairs of automobiles, then we have many dissimilarity matrices, one for each driver.

What is the MDS model? Each object or event is represented by a point in a multidimensional space. The points are arranged in this space so that the distances between pairs of points have the strongest possible relation to the similarities among pairs of objects. That is, two similar objects are represented by two points that are close together, and two dissimilar objects are represented by two points that are far apart. The space is usually a two- or three-dimensional Euclidean space but may have more dimensions. This model is called the Euclidean model. Sometimes, a model that was originally designed to portray individual differences in judgments of dissimilarity is used. This model, which is called the INDSCAL (individual differences scaling) model, uses weights for individuals on the dimensions of the Euclidean space. This model is discussed in greater detail below.

What types of MDS analyses are there? MDS is a generic term that includes many different types of analyses. The type of MDS analysis depends on the number of dissimilarity matrices, the measurement level of the dissimilarity data, and the MDS model used to analyze the data.

We can classify the types of MDS analyses according to how many matrices there are and what model is being used in the analysis. This gives us four kinds of MDS analyses: classical MDS (one matrix, Euclidean model), replicated MDS (several matrices, Euclidean model), weighted MDS (several matrices, weighted Euclidean model—also called the INDSCAL model), and generalized MDS (several matrices, general Euclidean model). For the INDSCAL and generalized models, you need more than one data matrix.

We can also classify types of MDS according to whether the dissimilarities data are measured on an ordinal scale (called **nonmetric MDS**) or an interval or ratio scale (**metric MDS**). The nonmetric/metric distinction can be combined with the classical/replicated/weighted distinction to provide six different types of MDS analyses. We discuss these types in the following sections.

The SPSS multidimensional scaling algorithm is capable of many kinds of analyses, not all of which are covered in this chapter. See Young and Lewyckyj (1979) or Young and Hamer (1987) for more information.

Example: Flying Mileages

The purpose of MDS is to construct a map of the locations of objects relative to each other from data that specify how different the objects are. This is similar to the problem faced by a surveyor who, once he has surveyed the distances between a set of places, needs to draw a map showing the relative locations of those places.

Consider a road map, which shows cities and the roads between them. On such a map, there is usually a small table of driving mileages between major cities on the map. In our example, we start with such a table, except that it is a table of flying mileages rather than driving mileages. With MDS, we can use these data to construct a map showing where the cities are located, relative to each other. We use these data because most of us are familiar with the relative locations of the large cities in the United States.

We know what structure to expect, and we will be looking to see if MDS can show us that structure.

In Figure 14-1, we present a matrix of dissimilarity data entered using the SPSS Data Editor. These data are actually the flying mileages between 10 American cities. The cities are the "objects" and the mileages are the "dissimilarities." For example, the distance between Atlanta and Chicago (shown in row 2, column 1) is 587 miles. Chicago and Denver are 920 miles apart, as shown in row 3, column 2. Note that all diagonal values are 0, since they represent the distance between a city and itself. Note also that each column of the matrix contains data for a different variable. The first column contains distances for *atlanta*, the second column contains *chicago* distances, and so forth.

Figure 14-1
Flying mileages between 10 American cities

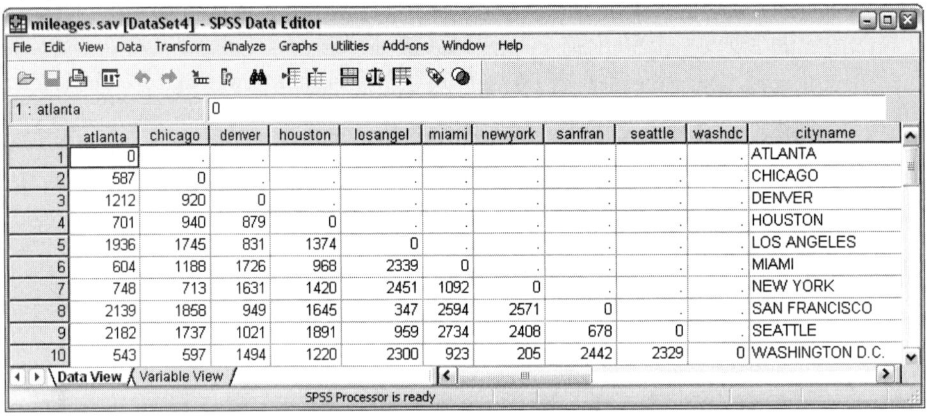

	atlanta	chicago	denver	houston	losangel	miami	newyork	sanfran	seattle	washdc	cityname
1	0	.	.	.	.	.	.	.	.	.	ATLANTA
2	587	0	.	.	.	.	.	.	.	.	CHICAGO
3	1212	920	0	.	.	.	.	.	.	.	DENVER
4	701	940	879	0	.	.	.	.	.	.	HOUSTON
5	1936	1745	831	1374	0	.	.	.	.	.	LOS ANGELES
6	604	1188	1726	968	2339	0	.	.	.	.	MIAMI
7	748	713	1631	1420	2451	1092	0	.	.	.	NEW YORK
8	2139	1858	949	1645	347	2594	2571	0	.	.	SAN FRANCISCO
9	2182	1737	1021	1891	959	2734	2408	678	0	.	SEATTLE
10	543	597	1494	1220	2300	923	205	2442	2329	0	WASHINGTON D.C.

To obtain an MDS analysis, open the data file *mileages.sav* and from the menus choose:

Analyze
 Scale
  Multidimensional Scaling (ALSCAL)...

▶ Variables: atlanta, chicago, denver, houston, losangel, miami, newyork, sanfran, seattle, washdc
Distances
⊙ Data are distances

Model...
  Level of Measurement
  ⊙ Ratio
  Scaling Model
  ⊙ Euclidean distance
  Conditionality
  ⊙ Matrix
  Dimensions
  Minimum: 2 Maximum: 2

Options...
  Display:
  ☑ Group plots
  ☑ Data matrix
  Criteria
  Minimum s-stress value: .001

Figure 14-2 shows the MDS map based on these data; it shows the relative locations of the 10 cities in the United States. The plot has 10 points, one for each city.

Figure 14-2
MDS plot of intercity flying mileage

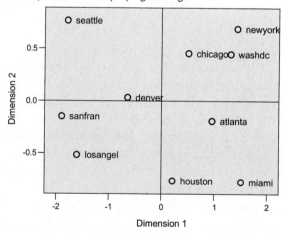

Cities that are similar (have short flying mileages) are represented by points that are close together, and cities that are dissimilar (have long mileages) are represented by points that are far apart. Note that the orientation is arbitrary. For this particular analysis, SPSS has determined the dimensions such that the first dimension is the longest, the second is the next longest, and so forth. By coincidence, north is located at the top (rather than the bottom), and west to the left (instead of right).

Nature of Data Analyzed in MDS

The SPSS Multidimensional Scaling procedure analyzes dissimilarity data. The data can either be obtained directly in some empirical situation or computed from multivariate data. In either case, the basic nature of dissimilarity data is exemplified by the flying mileages: the individual elements of the data matrix indicate the degree of dissimilarity between the pairs represented by the rows and columns of the matrix. In the flying mileage example, the dissimilarities are the various flying mileages and the pairs of objects are pairs of cities.

SPSS is very flexible about the kinds of dissimilarity data that can be analyzed. The data can vary in terms of their measurement level, shape, and conditionality. There can be any pattern of missing data. These aspects are discussed briefly in this section. For a more detailed discussion, see Young and Hamer (1987).

The SPSS Multidimensional Scaling procedure can directly analyze multivariate data, but we don't usually recommend that it be used for this purpose. SPSS can calculate dissimilarities from multivariate data, which can then be used by the scaling algorithm.

Measurement Level of Data

The data can be analyzed in ways that respect the basic nature of any of three commonly identified levels of measurement: ordinal, interval, or ratio. Examples of nominal dissimilarities data are exceedingly rare and are not discussed here. The remaining three levels of measurement (ordinal, interval, and ratio) fall into two categories that define two types of data and two types of analyses. The ordinal level of measurement defines data that are qualitative and analyses that are nonmetric. The interval and ratio levels, on the other hand, define data that are quantitative and analyses that are metric. All three measurement levels, and both metric and nonmetric analyses, will be discussed in this chapter.

Shape of Data

A less commonly discussed aspect of data is their shape. Data can be square or rectangular. Square data, in turn, can be **symmetric** or **asymmetric**. Thus, we have three basic data shapes: symmetric, asymmetric, and rectangular. SPSS can analyze all three shapes.

In this chapter, we discuss only square data, both symmetric and asymmetric. For both types of data, the rows and columns refer to the same set of objects. For example, the rows and columns of square data may refer to political candidates or to automobiles. In both examples, the data indicate the degree of dissimilarity of a pair from the set of objects. Thus, if the rows and columns refer to automobiles, the data indicate the degree of dissimilarity of pairs of automobiles.

The difference between symmetric and asymmetric data is whether the degree of dissimilarity between, say, a Ford and a Chevrolet is the same as the dissimilarity between a Chevrolet and a Ford, regardless of the order in which the two objects are considered. If the dissimilarity is the same, the data are symmetric. Otherwise, they are asymmetric.

The SPSS Multidimensional Scaling procedure can also analyze rectangular data. The most common example of rectangular data is multivariate data. Rectangular data elements specify the dissimilarity of all objects in one set to all objects in a second set. But for these data, there is no information about the dissimilarity of objects within either set, which is often a problem for MDS. Rectangular data should be analyzed directly with the greatest care, since the results are often not robust.

Conditionality of Data

In this chapter, we discuss two kinds of data conditionality: matrix conditionality and row conditionality. Most dissimilarity data are **matrix conditional**. That is, most dissimilarities are such that the numbers within a matrix are on the same measurement scale. Thus, if we have a matrix where the dissimilarity index is the proportion of times that a pair of senators voted differently out of all their votes during a session of the senate, all of the proportion indexes are on the same scale and the data are said to be matrix conditional. As a second example, consider judgments of dissimilarity about pairs of wines made by several different experts. Here, the judgments for a given expert are presumed to be on a single scale, but the experts probably each have their own idiosyncratic ways of responding, so the actual meaning of a specific judgment is conditional on which expert made the judgment. Thus, these data are also matrix conditional. If the ratings were of individual wines, with each judge's ratings in a row and each wine in a separate column of a rectangular matrix, the ratings would be row conditional.

Missing Data

SPSS can analyze data with any pattern of missing values. Of course, the more values that are missing from the data, the lower the probable reliability of the analysis. You should always try to have data with as few missing values as possible. Certain patterns of missing data are not susceptible to robust analysis, as discussed by Young and Hamer (1987).

Multivariate Data

As has been discussed by Young and Hamer (1987), multivariate data can be viewed as rectangular (since the number of observations usually does not equal the number of variables) and column conditional (since the variables are usually measured on different scales). The measurement level of multivariate data often varies from variable to variable, some variables being nominal, others ordinal, and still others interval.

While SPSS can analyze multivariate data directly, it rarely produces useful results. If the multivariate data are quantitative (that is, if all of the variables are at the interval or ratio levels of measurement), or if all variables are dichotomous, SPSS can calculate dissimilarity data from the multivariate data, using Euclidean or binary Euclidean distances. These dissimilarities can then be analyzed by SPSS.

While the SPSS Multidimensional Scaling procedure can be used to analyze a matrix of dissimilarities calculated from a multivariate matrix, there is no real advantage in doing so if you have only one multivariate matrix. The analysis will be equivalent to a principal components analysis when the default specifications are used, and the SPSS Factor Analysis procedure can perform the same analysis more efficiently and easily.

The real strength of SPSS with multivariate data is when you have multiple multivariate matrices. In this case, you can calculate a separate dissimilarity matrix for each multivariate matrix. These multivariate matrices can then be simultaneously analyzed by SPSS. This feature provides you with three-mode factor analysis, a family of analyses that provide one of the few methods for analyzing multiple-matrix multivariate data in any statistical system.

Classical MDS

Classical MDS (CMDS) is the simplest kind of MDS. The identifying aspect of CMDS is that there is only one dissimilarity matrix. SPSS can analyze many dissimilarity matrices simultaneously, resulting in other kinds of MDS that are not CMDS. They are discussed later in this chapter.

Example: Flying Mileages Revisited

Consider the flying mileage example at the beginning of the chapter. This is an example of classical MDS because there is only one dissimilarity matrix, the matrix of flying mileages shown in Figure 14-1. The data are square symmetric and are at the ratio level of measurement. The only model that can be used for a single matrix of dissimilarities is the Euclidean model, and we know that two dimensions are appropriate.

Figure 14-3 summarizes all options in effect for this particular MDS analysis. Figure 14-4 gives the data matrix, which was requested for this analysis. The mileage values are identical to those shown in Figure 14-1.

Figure 14-3
Analysis options

```
Alscal Procedure Options


Data Options-

Number of Rows (Observations/Matrix).    10
Number of Columns (Variables) .  .  .    10
Number of Matrices  .  .  .  .  .  .     1
Measurement Level .  .  .  .  .  .  .    Ratio
Data Matrix Shape .  .  .  .  .  .  .    Symmetric
Type  .  .  .  .  .  .  .  .  .  .  .    Dissimilarity
Approach to Ties  .  .  .  .  .  .  .    Leave Tied
Conditionality .  .  .  .  .  .  .  .    Matrix
Data Cutoff at .  .  .  .  .  .  .  .     .000000


Model Options-

Model .  .  .  .  .  .  .  .  .  .  .    Euclid
Maximum Dimensionality  .  .  .  .  .    2
Minimum Dimensionality  .  .  .  .  .    2
Negative Weights  .  .  .  .  .  .  .    Not Permitted


Output Options-

Job Option Header .  .  .  .  .  .  .    Printed
Data Matrices  .  .  .  .  .  .  .  .    Printed
Configurations and Transformations .    Plotted
Output Dataset .  .  .  .  .  .  .  .    Not Created
Initial Stimulus Coordinates .  .  .    Computed


Algorithmic Options-

Maximum Iterations  .  .  .  .  .  .       30
Convergence Criterion  .  .  .  .  .      .00100
Minimum S-stress  .  .  .  .  .  .  .     .00100
Missing Data Estimated by  .  .  .  .    Ulbounds
```

Figure 14-4
Matrix of flying mileages

```
        Raw (unscaled) Data for Subject 1

             1         2         3         4         5

  1       .000
  2    587.000      .000
  3   1212.000    920.000      .000
  4    701.000    940.000    879.000      .000
  5   1936.000   1745.000    831.000   1374.000      .000
  6    604.000   1188.000   1726.000    968.000   2339.000
  7    748.000    713.000   1631.000   1420.000   2451.000
  8   2139.000   1858.000    949.000   1645.000    347.000
  9   2182.000   1737.000   1021.000   1891.000    959.000
 10    543.000    597.000   1494.000   1220.000   2300.000

             6         7         8         9        10

  6       .000
  7   1092.000      .000
  8   2594.000   2571.000      .000
  9   2734.000   2408.000    678.000      .000
 10    923.000    205.000   2442.000   2329.000      .000
```

The iteration history in Figure 14-5 is produced by default. However, if a minimum S-stress of 0.001 is not specified, SPSS produces only one iteration, since the default minimum value for S-stress for further iterations is 0.005, and the S-stress value for the first iteration is only 0.003. **S-stress** is a measure of fit ranging from 1 (worst possible fit) to 0 (perfect fit). Next, the procedure displays two more measures of fit: Kruskal's stress measure (0.002) and the squared correlation coefficient (R-square $= 1.000$) between the data and the distances. All three measures of fit, which are discussed and defined below, indicate that the two-dimensional Euclidean model describes these flying mileages perfectly. The stimulus coordinates in Figure 14-6 are the coordinates used to generate the plot in Figure 14-2.

Figure 14-5
Iteration history

```
Iteration history for the 2 dimensional solution (in squared distances)

              Young's S-stress formula 1 is used.

        Iteration      S-stress      Improvement

            1           .00308
            2           .00280         .00029

              Iterations stopped because
        S-stress improvement is less than    .001000


        Stress and squared correlation (RSQ) in distances

RSQ values are the proportion of variance of the scaled data (disparities)
        in the partition (row, matrix, or entire data) which
        is accounted for by their corresponding distances.
        Stress values are Kruskal's stress formula 1.

              For  matrix
    Stress  =   .00231      RSQ =  .99998
```

Figure 14-6
Stimulus coordinates

```
Configuration derived in 2 dimensions

              Stimulus Coordinates

                   Dimension

Stimulus    Stimulus       1          2
Number       Name
    1        atlanta      .9587     -.1913
    2        chicago      .5095      .4537
    3        denver      -.6435      .0330
    4        houston      .2150     -.7627
    5        losangel   -1.6043     -.5159
    6        miami       1.5104     -.7733
    7        newyork     1.4293      .6906
    8        sanfran    -1.8940     -.1482
    9        seattle    -1.7870      .7677
   10        washdc      1.3059      .4464
```

Next, the procedure displays a matrix called *Optimally scaled data (disparities) for subject 1*, as shown in Figure 14-7. For this analysis, in which the data are specified to be at the ratio level of measurement, the values in this matrix of disparities are linearly related to the original flying mileages.

Figure 14-7
Optimally scaled data for subject 1

```
Optimally scaled data (disparities) for subject    1

            1         2         3         4         5

 1        .000
 2        .783      .000
 3       1.616     1.227      .000
 4        .935     1.254     1.172      .000
 5       2.582     2.327     1.108     1.832      .000
 6        .806     1.584     2.302     1.291     3.119
 7        .998      .951     2.175     1.894     3.269
 8       2.853     2.478     1.266     2.194      .463
 9       2.910     2.317     1.362     2.522     1.279
10        .724      .796     1.992     1.627     3.067

            6         7         8         9        10

 6        .000
 7       1.456      .000
 8       3.459     3.429      .000
 9       3.646     3.211      .904      .000
10       1.231      .273     3.257     3.106      .000
```

The perfect fit summarized by the three fit indexes is represented in Figure 14-8. This plot is a scatterplot of the raw data (horizontal axis) versus the distances (vertical axis). The raw data have been standardized, so their units have been changed. The distances are the Euclidean distances between all pairs of points shown in the configuration plot. This plot is an ordinary scatterplot and is interpreted as such. It represents the fit of the distances to the data, which is the fit that is being optimized by the procedure and that is summarized by the fit indexes. In fact, *RSQ* is simply the squared correlation between the data and the distances. Thus, we look at this scatterplot to see how much scatter there is around the perfect-fit line that runs from the lower left to the upper right. In this analysis, we see that there is no scatter and no departure from perfect linear fit.

Figure 14-8
Scatterplot of raw data versus distances

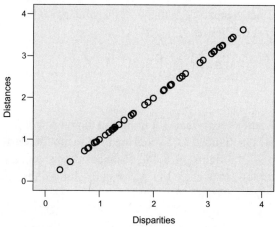

These high-quality results occur because the data have essentially no error, and because we have properly assumed that the data are at the ratio level of measurement. These results also imply that a two-dimensional space is sufficient to explain the flying mileages between cities in the United States. These results, then, assure us that the scaling algorithm is doing what it is supposed to do, since it is recovering structure that we know to be in the data.

Euclidean Model

In this section, we present the **Euclidean model**. First, we present the algebra of the model and then the geometry. Then we return to the algebra of the model, this time presented in matrix algebra instead of scalar algebra. Since the details of the matrix algebra are not crucial to the remainder of the chapter, they can be skipped.

Algebra of the Euclidean Model

Classical MDS employs Euclidean distance to model dissimilarity. That is, the distance d_{ij} between points i and j is defined as

$$d_{ij} = \left[\sum_{a}^{r} (x_{ia} - x_{ja})^2 \right]^{1/2}$$

where x_{ia} specifies the position (coordinate) of point i on dimension a. The x_{ia} and x_{ja} are the stimulus coordinates in Figure 14-6, which are used to plot the derived stimulus configuration in Figure 14-2. In Figure 14-8, the distances d_{ij} are plotted on the vertical axis of the scatterplot versus the corresponding dissimilarities (after normalization) on the horizontal axis.

Note that there are n points in Figure 14-2. In this case, $n = 10$, one for each of the n objects (cities). There are also r dimensions (here, $r = 2$). The default value of r is 2, but you can specify a different number of dimensions. The dissimilarity of objects i and j is denoted as s_{ij}, displayed in Figure 14-4 under the heading *Raw (unscaled) Data for Subject 1*. In this case, the dissimilarities are the actual flying mileages.

The dimensions of the space can be reflected, translated, permuted, and rotated, and they can all be rescaled by the same scaling factor. They cannot, however, be individually rescaled by different scaling factors.

Geometry of the Euclidean Model

Geometrically, the Euclidean distance model presented above is a multidimensional generalization of the two-dimensional Pythagorean theorem, as demonstrated in Figure 14-9.

Figure 14-9
Euclidean distance (not drawn in SPSS)

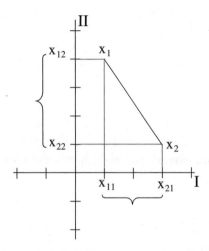

In this figure, the points x_1 and x_2 are shown with dashed lines that project orthogonally onto the two dimensions. The place that these dashed lines project onto the dimensions corresponds to the two coordinates for each point. Thus, the point x_1 has projection x_{11} on dimension 1 and x_{12} on dimension 2, while the point x_2 has projections x_{21} and x_{22} on the two dimensions. By comparing this figure with the previous formula, we see that the formula calculates d_{12} (the length of the hypotenuse of the right triangle) as the square root of the sum of the squared lengths of the two other sides of the triangle, where these squared lengths are $(x_{11} - x_{21})^2$ and $(x_{12} - x_{22})^2$, respectively.

Matrix Algebra of the Euclidean Model

Using matrix algebra, the Euclidean model can be defined as

$$d_{ij} = [(x_i - x_j)(x_i - x_j)']^{1/2}$$

where x_i is the *i*th row of *X*, consisting of the *r* coordinates of the *i*th point on all *r* dimensions. The dimensions of a Euclidean space can be rotated. This characteristic will be contrasted with other MDS models that are not rotatable.

Normalization. The stimulus coordinates X can be changed in a number of ways that do not change the distances by more than a multiplicative constant (the type of change in scaling of the distances that is permitted by the Euclidean model). Permissible changes in the coordinates include changing the mean of each dimension and multiplying all coordinates by a constant. Since the mean and unit are arbitrary, these are fixed to be the same for all analyses. In particular, the mean of each dimension is made to be 0:

$$\sum_{i}^{n} x_{ia} = 0$$

The dimensions are given an average length equal to the number of points:

$$\sum_{i}^{n}\sum_{a}^{r} x_{ia}^{2} = nr$$

Note that nothing is done to rotate the space to any specific orientation. However, the first iteration is begun with the configuration oriented so that the first dimension is the longest, the second the next longest, and so forth ("longest" means greatest sum of squared coordinates).

Details of CMDS

In this section, we discuss some important details of CMDS, including the distinction between metric and nonmetric CMDS, the measures of goodness of fit displayed by SPSS, and the fundamental CMDS equation.

Metric CMDS

The seminal MDS work was done by Torgerson (1952) and is now known as **metric CMDS**. The flying mileage example presented above is an example. It is *metric* because the flying mileages are assumed to be at the ratio level of measurement. It is *classical* because there is exactly one matrix of dissimilarities. For metric classical MDS, SPSS fits the squared Euclidean distances D^2 to the dissimilarities S so that they are as much like S as possible, in a least-squares sense. (The matrix D^2 has elements that are the squares of the elements of the matrix D.)

The fitting of the squared Euclidean distances to the dissimilarities is represented by the equation:

$$l\{S\} = D^2 + E$$

where $l\{S\}$ is read "a linear transformation l of the dissimilarities S." The transformation l takes the matrix S as its argument and yields as its value the matrix:

$$T = l\{S\}$$

If the measurement level is ratio (as in the example), then the linear transformation has a 0 intercept. The intercept can be non-zero when the level is interval. For dissimilarities, the slope of the transformation is positive. If the data are row conditional, the transformations are somewhat more complex, as will be explained below.

Note that the transformations for the interval measurement level are subject to indeterminacies that do not occur at other measurement levels. If warning messages appear, you should try another measurement level.

Note also that, for technical reasons, similarities are not as robust as dissimilarities for the SPSS Multidimensional Scaling procedure, so you should convert similarities to dissimilarities before performing multidimensional scaling. This recommendation holds regardless of other analysis details.

In the preceding equation, E is a matrix of errors (residuals). SPSS minimizes S-stress, a normalized sum of squares of this matrix, as described below. Thus, the S-stress measure is defined on the squared distances in D^2, whereas the stress and R measures are defined on the (unsquared) distances in D. This point has certain implications for information output by the procedure, as will be discussed in the examples below.

Since the distances D (and the squared distances D^2) are a function of the coordinates X, the goal of CMDS is to calculate the coordinates X so that the sum of squares of E is minimized, subject to suitable normalizations. SPSS also calculates the slope (and intercept) of the transformation $l\{S\}$ to minimize E.

Nonmetric CMDS

The first major breakthrough in multidimensional scaling was due to Shepard (1962) and Kruskal (1964) and is known as **nonmetric CMDS**. Nonmetric CMDS refers to analyses in which the measurement level is specified to be ordinal.

For nonmetric CMDS, we define

$$m\{S\} = D^2 + E$$

where $m\{S\}$ is "a monotonic transformation *m* of the dissimilarities *S*." As with the transformation *l*, *m* takes the matrix *S* as its argument, and yields as its value the matrix:

$$T = m\{S\}$$

If *S* is actually dissimilarities, then $m\{S\}$ preserves rank order, whereas if *S* is actually similarities, then $m\{S\}$ reverses rank order. Thus, for nonmetric CMDS, SPSS solves for the monotonic (order-preserving) transformation $m\{S\}$ and the coordinates *X*, which together minimize the sum of squares of the errors *E* (after normalization).

The matrix of transformed dissimilarities $m\{S\}$ (or the matrix $l\{S\}$) appears in output under the heading *Optimally scaled data (disparities) for subject 1*, as shown in Figure 14-7. This terminology refers to the fact that the transformation *m* (or *l*) induces a new scaling of the data that is the scaling which optimizes the S-stress index. It also refers to the fact that, historically, this information was originally called "disparities."

The nonmetric minimization problem represents a much more difficult problem to solve than the corresponding metric problem. The nonmetric minimization problem belongs to the general class of problems discussed by Young (1981). These problems require iterative solutions, implying that nonmetric analyses take much more computer time than the metric ones.

Measures of Fit

As shown in Figure 14-5, SPSS produces an iteration history that lists, for each iteration, the S-stress and improvement of the iteration. Note that the fit formula is based on the squared distances contained in the matrix D^2. This is the S-stress index proposed by Takane et al. (1977). The improvement simply represents the amount of improvement in S-stress from one iteration to the next. S-stress indicates the fit of the squared distances D^2 to the transformed data $m\{S\}$ or $l\{S\}$.

In the presentation above, we denoted the result of the monotonic and linear transformations of the data by the matrix *T*, so that $T = m\{S\}$ or $T = l\{S\}$, depending on measurement level. Then,

$$S\text{-stress} = \left(\frac{\|E\|}{\|T\|} \right)^{1/2}$$

where the notation $\|E\|$ indicates the sum of all squared elements of the error (residual) matrix E, and $\|T\|$ indicates the sum of all squared elements of the matrix of transformed data T. Note that

$$\|E\| = \|T - D^2\|$$

which means that S-stress is the square root of the ratio of the error sums of squares to the total sums of squares, where the error sums of squares is calculated between the squared distances and the transformed data, and the total sums of squares is calculated on the transformed data. That is, S-stress is the square root of the proportion of the total sums of squares of the transformed data which is error. Error is indicated by lack of fit of the squared distances to the transformed data.

Except for the S-stress index, all results (including the two additional measures of fit) are reported in terms of the distances D, not squared distances D^2. Two additional indexes of fit are stress and RSQ. The **stress index** is Kruskal's (1964) stress formula. It is defined in exactly the same fashion as S-stress, except that distances are used instead of squared distances. Note that the SPSS multidimensional scaling algorithm optimizes S-stress, not stress. This has certain implications for information output by the procedure, which will be discussed in the examples below.

RSQ is the squared simple correlation between corresponding elements of T and D. It can be interpreted as the proportion of variance of the transformed data T that is accounted for by the distances D of the MDS model.

Fundamental CMDS Equation

We can now summarize, in a single equation, the CMDS data analysis problem being solved by the MDS procedure. The fundamental equation is

$$S \overset{t}{=} T = D^2 + E$$

which reads S (the original dissimilarity data) are equal, by transformation (the symbol $\overset{t}{=}$ represents the error-free one-to-one transformation), to T (the transformed dissimilarity data), which in turn are equal to the model's squared Euclidean distances D^2 plus error E. SPSS solves for D^2 and for $\overset{t}{=}$ so that the sum of the squared elements of E is minimized.

We can think of the transformed data *T* as an interface between the measurement characteristics of *S* (which may have a variety of measurement levels, shapes, and conditionalities, and which may have missing values) and those of D^2 (which is always at the ratio level of measurement, is always symmetric and unconditional, and never has missing values). *T* is, in fact, identical to *S* when looked at through the filters imposed by *S*'s various measurement characteristics. That is, *T* and *S* have identical ordinal characteristics if the measurement level of *S* is ordinal. Similarly, *T* is identical to $D^2 + E$ when viewed through their measurement characteristic's filters.

Example: Ranked Flying Mileages

As an example of nonmetric CMDS, we now return to the flying mileages, but we use their ranks instead of the actual mileages. We still hope that the scaling algorithm can construct a reasonable map of the United States, even though we have degraded the information from mileages to ranks.

Open the data file *rankedmileages.sav* and from the menus choose:

Analyze
  Scale
    Multidimensional Scaling (ALSCAL)...

    ▶ Variables: atlanta, chicago, denver, houston, losangel, miami, newyork, sanfran, seattle, washdc
    Distances
    ⊙ Data are distances

  Model...
    Level of Measurement
    ⊙ Ordinal
    Scaling Model
    ⊙ Euclidean distance
    Conditionality
    ⊙ Matrix
    Dimensions
    Minimum: 2 Maximum: 2

  Options...
    Display:
    ☑ Group plots
    ☑ Data matrix
    Criteria
  Minimum s-stress value: .001

In Figure 14-10, we show the new matrix of dissimilarity data for intercity flying mileages. These data differ from the first set of data in that they have been converted into their ranking numbers—that is, an element in this new data matrix specifies the rank position of the corresponding mileage element in the first data matrix. Note that there are 45 elements and that the ranks range from 1 through 45. Thus, the flying mileage between New York and Washington, D.C., is the shortest (rank of 1), while the flying mileage between Miami and Seattle is the longest (45). Since these data are at the ordinal level of measurement, they are suitable for nonmetric CMDS. A nonmetric CMDS of these data produces the plot in Figure 14-11, which is virtually indistinguishable from the plot in Figure 14-2.

Figure 14-10
Ranked distances between cities

rankedmileages.sav [DataSet5] - SPSS Data Editor										
File Edit View Data Transform Analyze Graphs Utilities Add-ons Window Help										

1 : atlanta   0

	atlanta	chicago	denver	houston	losangel	miami	newyork	sanfran	seattle	washdc
1	0	.	.	.	.	.	.	.	.	.
2	4	0	.	.	.	.	.	.	.	.
3	22	13	0	.	.	.	.	.	.	.
4	8	15	12	0	.	.	.	.	.	.
5	34	31	11	24	0	.	.	.	.	.
6	6	21	29	18	39	0	.	.	.	.
7	10	9	27	25	42	20	0	.	.	.
8	35	32	16	28	2	44	43	0	.	.
9	36	30	19	33	17	45	40	7	0	.
10	3	5	26	23	37	14	1	41	38	0

Data View / Variable View /

SPSS Processor is ready

Figure 14-11
Plot of nonmetric (ranked) distances

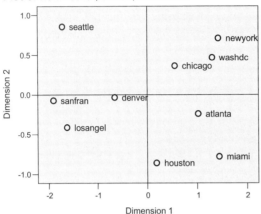

The remaining output for this analysis is shown in Figure 14-12 through Figure 14-17. SPSS displays a warning message, telling us that the number of parameters being estimated (which for these data is 20: $n = 10 \times r = 2$) is dangerously large compared to the number of elements in the data matrix (which is $n(n-1)/2 = 45$). This warning message did not appear for the previous analysis because we had assumed the data were quantitative. In general, we should be cautious when this message appears.

Next, the procedure displays the iteration history for the two-dimensional solution, shown in Figure 14-12. Note that the S-stress index starts at 0.03346, and after eight iterations reduces to 0.00585, where it stops due to the default minimum improvement criterion of 0.001 being reached. This index of fit is not as good as in the metric CMDS of the flying mileages, reflecting the fact that the precision of the data has been reduced because ranks are used instead of actual mileages.

Figure 14-12
Iteration history

```
Iteration history for the 2 dimensional solution(in squared distances)

            Young's S-stress formula 1 is used.

        Iteration      S-stress      Improvement

            1           .03346
            2           .02259         .01086
            3           .01656         .00604
            4           .01270         .00385
            5           .01009         .00261
            6           .00821         .00188
            7           .00684         .00137
            8           .00585         .00100
               Iterations stopped because
        S-stress improvement less than   .001000
```

Next, the procedure gives the two additional measures of fit—stress (0.008) and RSQ (1.000)—as shown in Figure 14-13. The stress measure indicates worse fit than in the metric analysis, as would RSQ if more decimal places were displayed. (The RSQ value has been rounded *up* to 1.) Even though the fit has become somewhat worse, the overall conclusion is that the squared distances D^2 of this two-dimensional Euclidean model describe the monotonic transformation of the data $T = m\{S\}$ perfectly.

Figure 14-13
Measures of fit

```
Stress and squared correlation (RSQ) in distances

RSQ values are the proportion of variance of the scaled data (disparities)
            in the partition (row, matrix, or entire data) which
            is accounted for by their corresponding distances.
            Stress values are Kruskal's stress formula 1.

                  For   matrix
        Stress  =    .00782      RSQ =  .99968
```

Figure 14-14 shows the stimulus coordinates, which are very similar to the coordinates from the metric analysis in Figure 14-6.

Figure 14-14
Stimulus coordinates

```
Configuration derived in 2 dimensions

                        Stimulus Coordinates

                            Dimension

Stimulus    Stimulus       1         2
 Number       Name

    1        atlanta     1.0195    -.2397
    2        chicago      .5557     .3622
    3        denver      -.6609    -.0344
    4        houston      .1816    -.8583
    5        losangel   -1.6307    -.4110
    6        miami       1.4377    -.7782
    7        newyork     1.4227     .7106
    8        sanfran    -1.9047    -.0734
    9        seattle    -1.7195     .8553
   10        washdc      1.2987     .4668
```

Four plots are produced for this nonmetric analysis, whereas only two plots were produced for the metric analysis. For both analyses, the first two plots are the plot of the derived configuration (Figure 14-11) and the scatterplot of linear fit.

The scatterplot of linear fit, shown in Figure 14-15, plots the monotonically transformed data (disparities) $T = m\{S\}$ horizontally versus the distances D vertically. This plot displays the departures from linearity that are measured by the stress and RSQ indexes. We see that there is very little scatter, although what scatter there is appears for the small distances and disparities. For this nearly perfectly fitting analysis, there are some departures from perfect fit for the small transformed data but very few for the large transformed data. This is a common result and is an artifact of the fit index S-stress, which tends to overfit large data values and underfit small ones because the *squared* distances are being fit to the data. Figure 14-16 is the plot of nonlinear fit. It plots the raw data S horizontally versus the distances D vertically. This plot displays the same departures from fit as in the previous plot but displays them relative to the raw data instead of the transformed data. Here we see that large ranks are better fit than small ranks.

Figure 14-17 is the plot of transformation. It plots the raw data S horizontally versus the monotonically transformed data $T = m\{S\}$ vertically. The transformed data are called **disparities**. This plot displays the transformation that rescores the ranked flying mileages so that they are as much like the distances, given that the new scores must be in the same order as the original ranks. Since this is a monotonic transformation, the plotted values must never go down as we move from left to right (unless the data are similarities, in which case the transformation line moves monotonically from upper left to lower right). The transformation plot can be placed on top of the nonlinear fit plot and can be interpreted as the monotonic regression line that minimizes error, where error is measured vertically in both plots.

Figure 14-15
Scatterplot of linear fit

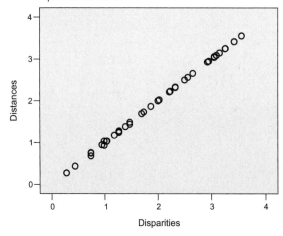

Figure 14-16
Plot of nonlinear fit

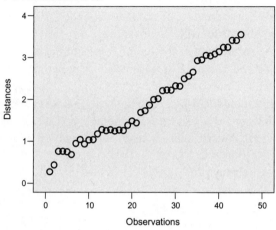

Figure 14-17
Plot of transformation

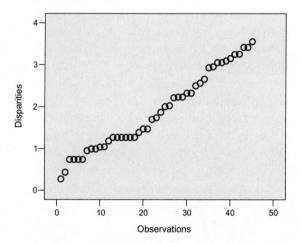

To interpret the transformation plot, we look at its overall shape and smoothness. This particular transformation is quite linear and smooth. The linearity leads us to conclude that the nonmetric analysis of the ranks is essentially the same as the metric analysis of the raw data. This conclusion would be stronger if the nonmetric analysis were of the actual flying mileages. Then, a linear transformation would yield the same results as the metric analysis. The smoothness suggests that we have a reasonably continuous, nondegenerate transformation. In particular, we should be suspicious of transformations that consist of a series of a few horizontal steps. These "step functions" suggest a discontinuous, possibly degenerate transformation.

Our overall conclusion, then, is that the results of the nonmetric analysis of the ranked flying mileages are very concordant with the results of the metric analysis of the flying mileages themselves. Since the transformation is quite linear in the nonmetric analysis, this tells us that the metric analysis, which is more parsimonious, is to be preferred. We can conclude from the two analyses that it is reasonable to assume that the flying mileages are in fact measured at the ratio level of measurement. We can also conclude from the metric analysis that the mileages are nearly error free and are basically two dimensional.

Repeated CMDS

All of the remaining types of MDS that we discuss differ from CMDS in that they are appropriate to data that consist of more than one matrix of dissimilarities. The major ways of analyzing such data are known as replicated MDS and weighted MDS.

It should be mentioned briefly that one way of analyzing multiple matrices of dissimilarity data is to repeatedly apply classical MDS, once to each matrix. This approach implies that you believe that the many matrices of data have no shared structure. The configurations of points underlying each dissimilarity matrix are presumed to be totally unrelated to each other in any fashion. If the several matrices of data are obtained from several individual judges, this is an individual differences model that permits the greatest freedom in modeling individual differences. There are, in fact, no constraints. However, this individual differences model is the least parsimonious model, since there is an entire configuration of points for each individual. If there are n points (cities, in the previous examples) with coordinates on r dimensions, there are $n \times r$ parameters per individual. If there are m individuals, there are $m \times n \times r$ parameters. Not only is this very nonparsimonious, but the results are difficult to interpret, since you are faced with the task of having to compare m separate analyses.

Replicated MDS

Historically, the next major development in the multidimensional scaling literature was **replicated MDS** (McGee, 1968). Replicated MDS (RMDS) was the first proposal that extended MDS to permit the analysis of more than one matrix of dissimilarities—a particularly important development, since researchers typically have more than one dissimilarity matrix. This is not the sole defining characteristic of RMDS, however, as the weighted MDS analyses discussed in later portions of this chapter also use several matrices of dissimilarities.

The defining characteristic of RMDS is that it applies the Euclidean distance model to several dissimilarity matrices simultaneously. The basic assumption is that the stimulus configuration X applies with equal validity to every matrix of data. Thus, the implication is that all of the data matrices are, except for error, the same. They are replicates of each other, since there are no systematic differences other than, perhaps, systematic response bias differences. This is the most parsimonious and most constrained model of individual differences in MDS. Note that the number of parameters is the same as in CMDS; there are a total of $n \times r$ parameters for the n points on r dimensions. Another distance model forms the foundation of weighted MDS. This model relaxes the assumption that all data matrices are replicates of each other. The model is less parsimonious and less constrained than RMDS.

For RMDS, the matrix of squared distances D^2 is still defined by the same Euclidean distance formula that is involved in CMDS. The difference is that the data consist of several dissimilarity matrices S_k, $k = 1, \ldots, m$, there being m data matrices in total. The analysis is such that the matrix of squared distances D^2 is calculated so that it is simultaneously like all of the several dissimilarity matrices S_k.

Details of RMDS

In this section, we discuss some details of RMDS, including the metric and nonmetric varieties, individual differences in response bias in RMDS, measures of fit in RMDS, and the fundamental RMDS equation.

Metric RMDS

For metric RMDS, the analysis is based on fitting the equation:

$$l_k\{S_k\} = D^2 + E_k$$

where $l_k\{S_k\}$ is the linear transformation of the kth dissimilarity matrix S_k that best fits the squared Euclidean distances D^2. The data may be similarities or dissimilarities and may be at the ratio or interval levels, just as in metric CMDS. The analysis minimizes the sum of the squared elements, where the sum is taken over all elements in all matrices E_k, subject to normalization of X. The details of the fit index and normalization are discussed later in this chapter.

Nonmetric RMDS

For nonmetric RMDS, SPSS minimizes the several E_k, just as in metric RMDS, except that the equation becomes

$$m_k\{S_k\} = D^2 + E_k$$

where $m_k\{S_k\}$ is the monotonic transformation of the dissimilarity matrix S_k that is least-squares fit to the squared distances in D^2. The data and their monotonic transformation are the same as for nonmetric CMDS, except that there are k matrices instead of just one.

Individual Response Bias Differences

It is important to notice that for RMDS each linear or monotonic transformation l_k or m_k is subscripted, letting each data matrix S_k have a unique linear or monotonic relation to the squared Euclidean distances contained in D^2. Since k ranges up to m, there are m separate linear or monotonic transformations, one for each of the m dissimilarity matrices S_k. This implies that RMDS treats all the matrices of the data as being related to each other (through D) by a systematic linear or monotonic transformation (in addition to the random error contained in E_k).

In psychological terms, RMDS accounts for differences in the ways subjects use the response scale (that is, differences in their response bias). Consider, for example, a response scale consisting of the numbers 1 through 9. Even though each subject uses the same response scale, he or she won't necessarily use the scale in the same way. It could be that one subject uses only categories 1, 5, and 9, while another uses only the even numbers, and a third uses only the middle three categories of 4, 5, and 6. These differences in response style (bias) are taken into account by the separate transformations for each subject.

Measures of Fit

As discussed above for CMDS, SPSS produces an iteration history that lists, for each iteration, the S-stress and improvement of the iteration. S-stress uses a formula based on the squared distances contained in the matrix D^2. To define the formula, we begin by denoting the matrix of transformed data by the symbol T_k, where $T_k = m_k\{S_k\}$ or $T_k = L_k\{S_k\}$, depending on measurement level. Then,

$$S\text{-stress} = \left[\frac{1}{m} \sum_{k}^{m} (\|E_k\| / \|T_k\|) \right]^{\frac{1}{2}}$$

where the notation $\|E_k\|$ indicates the sum of all squared elements of the error (residual) matrix E_k, and $\|T_k\|$ indicates the sum of all squared elements of the matrix of transformed data T_k. Note that

$$\|E_k\| = \|T_k - D^2\|$$

which means that for RMDS, S-stress is the square root of the mean of the ratio of the error sums of squares for a matrix to the total sums of squares for that matrix, where the error sums of squares is calculated between the squared distances and the transformed data for one matrix, and the total sums of squares is calculated on the transformed data for one matrix. That is, S-stress is the square root of the mean, averaged over the several matrices of data, of the proportion of the total sums of squares for one matrix of transformed data which is error. Error is indicated by lack of fit of the squared distances to the transformed data.

In addition, stress and RSQ (but not S-stress) are calculated and displayed for each matrix of data. These are exactly the same formulas as for CMDS, calculated separately for each matrix. Then, the average stress and RSQ (again, not S-stress) are calculated and displayed (labeled *Overall Stress and RSQ* in the output). (The average stress is the square root of the mean of the squared stress values, while the average RSQ is the mean of the RSQ values.) The stress indexes for each matrix are Kruskal's stress and can be interpreted as such. The RSQ for a matrix indicates the proportion of variance of the disparities in the matrix which is accounted for by the squared distances. The overall RSQ is an average squared correlation, and it indicates the average proportion of variance accounted for in all of the transformed data.

For row conditional data, the stress and RSQ (but not S-stress) values are displayed for each row of each matrix, as are the overall indexes for each matrix. The interpretation of these values corresponds to their interpretation for matrix conditional data.

Note: The final S-stress in the iteration list and the overall stress value do not correspond because the former is defined on squared distances and the latter on distances.

Fundamental RMDS Equation

We can now summarize, in a single equation, the RMDS data analysis problem being solved by the scaling procedure. The fundamental equation is

$$S_k \stackrel{\llcorner}{=} T_k = D^2 + E$$

This equation parallels the fundamental CMDS equation and is to be understood in a similar fashion.

Example: Perceived Body-Part Structure

Jacobowitz (see Young, 1974) used RMDS to study the way language develops as children grow to adulthood. In his experiment, he asked children and adults to judge the dissimilarity of all pairs of 15 parts of the human body. The judges were five-, seven-, and nine-year-olds, and adults. There were 15 judges at each age. Here we report two separate RMDS analyses, one for the seven-year-olds and the other for the adults. These data, and the RMDS of them, are discussed extensively by Young (1974).

Each of the 15 subjects in each of the four age groups was asked to do the following task. The experimenter wrote the 15 body-part terms on a slip of paper, 15 slips in all. The experimenter then selected one of the body-part terms and called it the "standard." He then asked the subject to select the term from the remaining 14 that seemed most similar to the standard. The experimenter then removed this slip and asked the subject to select the most similar term from the remaining 13. This continued until all 14 "comparison" terms were rank ordered by their similarity to the standard. This task was then repeated 15 times per subject, one time with each term as standard.

Table 14-1 shows the variable names and data for one subject. The value 0 indicates the "standard" term, the value 1 indicates the comparison term that was picked first (most similar), the value 2 is the term picked second, and so forth. Note that each row of the data contains the numbers 1 through 14, plus a 0. The numbers 1 through 14 indicate the judged similarity order, and the 0 indicates that a term was not compared to itself (that is, zero dissimilarity). In the complete data set (not shown) there are 225 rows of data, 15 sets (for 15 subjects) of 15 rows each (for 15 terms). Also note that each data matrix is asymmetric and that the data are **row conditional**, as the values in a row are ranked only relative to other values in the same row, not in other rows.

Table 14-1

Body-part data for one subject

Cheek	Face	Mouth	Head	Ear	Body	Arm	Elbow	Hand	Palm	Finger	Leg	Knee	Foot	Toe
0	1	2	3	4	9	7	10	6	5	8	12	11	13	14
2	0	1	3	4	8	6	10	12	5	11	7	9	13	14
2	1	0	3	4	6	8	12	7	10	5	9	11	13	14
3	1	2	0	4	5	6	7	10	8	9	12	11	13	14
3	2	4	1	0	5	9	10	7	8	6	11	12	13	14
3	2	6	1	8	0	7	10	12	9	14	4	11	5	13
13	11	12	10	14	4	0	1	2	9	3	5	7	6	8
11	12	13	10	14	5	1	0	2	9	3	6	4	7	8
13	8	10	11	14	9	3	5	0	1	2	12	7	4	6
12	11	7	13	14	10	6	4	1	0	2	8	9	3	5
10	11	6	8	7	5	9	14	1	2	0	12	13	4	3
13	9	12	10	14	4	5	11	6	7	8	0	1	2	3
13	12	11	10	14	9	6	2	5	8	7	1	0	3	4
13	11	14	10	12	6	7	9	4	8	5	2	3	0	1
13	12	14	10	11	5	8	9	6	7	3	2	4	1	0

To obtain the RMDS solution, open the data file *adultbody.sav* and from the menus choose:

Analyze
  Scale
    Multidimensional Scaling (ALSCAL)...

     ▶ Variables: cheek, face, mouth, head, ear, body, arm, elbow, hand, palm, finger, leg, knee, foot, toe
    Distances
    ⊙ Data are distances
    Shape: Square asymmetric

Model...
    Level of Measurement
    ⊙ Ordinal
    Conditionality
    ⊙ Row
    Dimensions
    Minimum: 1 Maximum: 3
    Scaling Model
    ⊙ Euclidean distance

Options...
    Display:
    ☑ Group plots
    ☑ Data matrix
    Criteria
Minimum s-stress value: .001

The RMDS results for adults and seven-year-olds are shown in Figure 14-18 and Figure 14-19. Both figures are based on plots generated by a scaling procedure and cannot be obtained from SPSS. (A two-dimensional MDS plot for the adult data is shown in Figure 14-23.)

Figure 14-18
Three-dimensional body-part RMDS for adults (not drawn in SPSS)

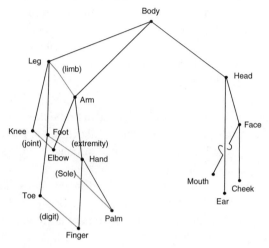

Figure 14-19
Three-dimensional body-part RMDS for seven-year-olds (not drawn in SPSS)

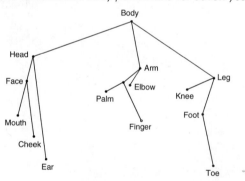

The lines were drawn to interpret the psycholinguistic structure that people have for body-part words. Jacobowitz theorized that the structure would be hierarchical. We can see that it is. (Note that there is nothing in the RMDS that forces the structure to be hierarchical. This structure is imposed on the solution by the data themselves, not by the analysis.) This hierarchical structure corresponds, psycholinguistically, to the phrase "has a." We say that a body "has a" head, which in turn "has a" face, which "has a" mouth.

Jacobowitz further theorized that the structure would become more complex as the children became adults. This theory is also supported, since the adults' hierarchy also involves a classification of corresponding arm and leg terms. In Figure 14-18, we have not only drawn the hierarchy, but we have also shown corresponding arm and leg terms linked by the dashed lines. In addition, the implied classification terms are shown in parentheses, and the term "sole," which was not a stimulus, is shown in the position that we would predict it to be in if the study were repeated with sole as a 16th stimulus.

The structure for adults shown in Figure 14-18 is based on a three-dimensional RMDS. For this analysis, there are 15 matrices of dissimilarity judgments. The judgments are specified as ordinal, asymmetric, and row conditional (conditionality and asymmetry are explained below). The default Euclidean model is used. When the Euclidean model is used with multiple matrices of dissimilarities, the result is RMDS. Three RMDS analyses are requested, the first with three dimensions, the second with two, and the last with one. Plots of results are also requested.

Figure 14-20 shows the iteration history for the three-dimensional solution for adults. The S-stress index starts at 0.24780 and improves only slightly to 0.23624 after six iterations. Iterations stop because the rate of improvement is slow. For this size data, the fit is rather good but not excellent.

It is not, unfortunately, possible to say exactly what a "good" or "bad" S-stress value is. We do know, however, that the value is a function of many things in addition to the amount of error in the data. For example, S-stress gets larger when the number of stimuli or matrices goes up.

Figure 14-20
Iteration history for the three-dimensional solution

```
Iteration history for the 3 dimensional solution(in squared distances)

            Young's S-stress formula 1 is used.

        Iteration     S-stress      Improvement

            1          .24780
            2          .24221         .00559
            3          .23959         .00262
            4          .23802         .00157
            5          .23697         .00105
            6          .23624         .00073
                 Iterations stopped because
        S-stress improvement less than   .001000
```

The next information displayed is a large number of fit indexes. The stress and RSQ indexes are displayed for every row of every matrix (there are $15 \times 15 = 225$ of these). In addition, an averaged stress and RSQ index is calculated for each matrix (subject). Then, an average RSQ (but not stress) index is displayed for each stimulus (body-part term). Finally, the RSQ and stress indexes are averaged over all data. It is useful to look over all of the fit measures to see if there are any that are particularly poor. Figure 14-21 shows the output for the first three subjects. For these data, there are no great anomalies, although we do note that subject 2 does not fit as well as subjects 1 and 3. Figure 14-22 shows the stimulus coordinates used to produce the plot of dimensions 1 and 2 in Figure 14-23. Figure 14-18 is based on the coordinates in Figure 14-22.

Figure 14-21
Fit indexes

```
Stress and squared correlation (RSQ) in distances

RSQ values are the proportion of variance of the scaled data (disparities)
          in the partition (row, matrix, or entire data) which
          is accounted for by their corresponding distances.
             Stress values are Kruskal's stress formula 1.

                          Matrix   1
        Stimulus    Stress    RSQ  Stimulus    Stress    RSQ

            1        .072     .974      2       .062     .982
            3        .092     .962      4       .043     .988
            5        .114     .937      6       .112     .876
            7        .077     .959      8       .120     .928
            9        .091     .954     10       .117     .907
           11        .211     .730     12       .114     .917
           13        .090     .958     14       .157     .882
           15        .141     .886

        Averaged (rms) over stimuli
        Stress  =   .115      RSQ =  .923


                          Matrix   2
        Stimulus    Stress    RSQ  Stimulus    Stress    RSQ

            1        .219     .764      2       .359     .403
            3        .033     .995      4       .316     .388
            5        .157     .880      6       .122     .853
            7        .241     .615      8       .210     .783
            9        .270     .602     10       .266     .528
           11        .032     .993     12       .207     .725
           13        .179     .836     14       .167     .868
           15        .159     .854

        Averaged (rms) over stimuli
        Stress  =   .215      RSQ =  .739

                          Matrix   3
        Stimulus    Stress    RSQ  Stimulus    Stress    RSQ

            1        .043     .991      2       .030     .996
            3        .091     .963      4       .020     .997
            5        .115     .936      6       .059     .966
            7        .103     .926      8       .198     .807
            9        .152     .872     10       .036     .991
           11        .149     .859     12       .071     .968
           13        .171     .850     14       .210     .793
           15        .137     .893

        Averaged (rms) over stimuli
        Stress  =   .122      RSQ =  .921
```

Figure 14-22
RMDS stimulus coordinates for adults

```
Configuration derived in 3 dimensions
                  Stimulus Coordinates

                              Dimension
Stimulus    Stimulus    1         2         3
Number      Name
   1        CHEEK     1.6756    -.8000     .2893
   2        FACE      1.7508    -.0705     .5369
   3        MOUTH     1.4671    -.5392     .8706
   4        HEAD      1.6199     .8363     .5339
   5        EAR       1.3866    -.7665    1.0340
   6        BODY       .6873    1.9263    -.4371
   7        ARM       -.4621     .7918   -1.1974
   8        ELBOW    -1.1132     .0736    -.9565
   9        HAND      -.5005    -.4627   -1.2429
  10        PALM      -.0353    -.9172   -1.5053
  11        FINGER    -.8682   -1.1875    -.3785
  12        LEG      -1.1363    1.1501     .5941
  13        KNEE     -1.5253     .4856     .7355
  14        FOOT     -1.4240    -.0328     .2423
  15        TOE      -1.5222    -.4873     .8812
```

Figure 14-23
Plot of RMDS stimulus coordinates for adults

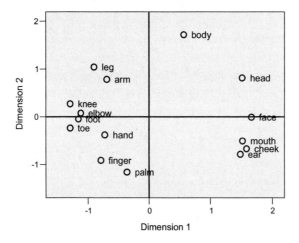

The scatterplot in Figure 14-24 is interpreted just like any other scatterplot, including the scatterplots discussed earlier in the analyses of the flying mileages and their ranks. The scatter is between the transformed data (horizontal axis, the disparities) and the distances between the points in the configuration (vertical axis). If the scatter is tight, the fit is good; if it is loose, it is bad. The overall RSQ value (0.846) is calculated from this scatterplot and is rather good.

Figure 14-24
Scatterplot of RMDS results for adults

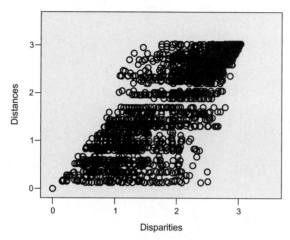

While the initial impression of this particular scatterplot is fairly poor, the density indicated by the plotting symbols reveals that the scatter is much more dense along the diagonal than away from it. Thus, the scatter is not as poor as it initially appears.

SPSS displays the same type of output for the two-dimensional solution and the one-dimensional solution. A notable feature of the three separate analyses is that the two-dimensional solution fits somewhat less well, but not a lot less well (S-stress = 0.27912) than the three-dimensional solution, whereas the one-dimensional solution fits quite a lot less well than the two- and three-dimensional solutions (S-stress = 0.35133). This suggests that the two-dimensional solution may be the best of these, since it is more parsimonious than the three-dimensional but fits nearly as well, and it fits quite a lot better then the one-dimensional solution.

Another notable feature of the three analyses is that the one-dimensional solution is clearly degenerate. The one-dimensional plot of the derived configuration in Figure 14-25 shows that the points are clustering together into two distinct clusters; on the bottom of the plot, we have all the head terms and body, whereas all the arm and leg terms are on top (body part labels are not shown in Figure 14-25). The cluster structure shows that the solution is degenerating into a simple geometric pattern, in this case consisting of all points in two places. Also, by looking at the scatterplot in Figure 14-26, we see a strange pattern that should be taken as another warning; there is a large horizontal gap in the scatter. This gap indicates that there are no medium-sized distances; instead, all distances are either large or small. This is another way of seeing the cluster-like nature of this solution.

Figure 14-25
Plot of one-dimensional solution

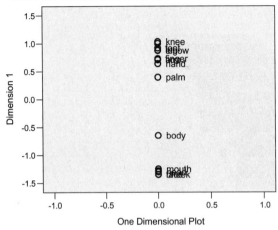

Figure 14-26
Scatterplot of one-dimensional solution

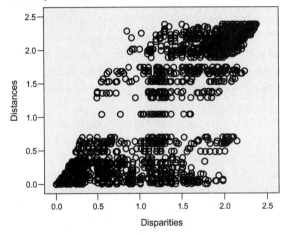

What we conclude from all these clues is that the one-dimensional analysis is actually telling us about the gross (as opposed to fine) structure of the data. The arm and leg terms are seen as being more similar to themselves than they are to the head terms or to the body terms. But the analysis appears to have degenerated into an oversimplified structure, suggesting in this case that we may have too few dimensions.

There are some indications that the two-dimensional solution may be partially degenerate. The two-dimensional plot of the derived configuration in Figure 14-27 looks quite circular, another kind of simple geometric pattern that should be interpreted cautiously and that may indicate that the analysis has degenerated into an oversimplified structure. We can see the hierarchical structure in this plot, so the analysis is, perhaps, only partially degenerate. However, caution is appropriate—this space may also be too low in dimensionality.

Figure 14-27
Plot of two-dimensional solution

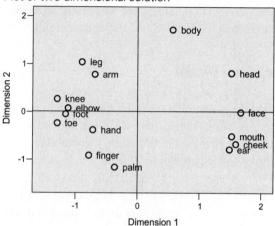

The conclusion that can be drawn from these three analyses is that the three-dimensional solution is the best. It fits reasonably well and shows no signs of degeneracy, whereas the two-dimensional and one-dimensional solutions seem somewhat questionable.

Weighted MDS

The second major breakthrough in multidimensional scaling was due to Carroll and Chang (1970). This development, which is called **weighted MDS (WMDS)**, generalized the Euclidean distance model so that the several dissimilarity matrices S_k could be assumed to differ from each other in systematically nonlinear or nonmonotonic ways. Whereas RMDS accounts only for individual differences in response bias, WMDS can also account for individual differences in the perceptual or cognitive processes that generate the responses. For this reason, WMDS is often called individual differences scaling (INDSCAL).

WMDS is based on the weighted Euclidean model. In this model, we still have the stimulus space *X* that we have had in the (unweighted) Euclidean model, but we also have a new weight space *W*. We can think of the stimulus space *X* as representing the information that is shared in common across the individuals about the structure of the stimuli, just as in RMDS. In addition, we can think of the weight space *W* as representing the information that is unique to each individual about the structure of the stimuli, a notion that we did not have in RMDS.

We now turn to a detailed presentation of WMDS. We discuss the geometry, algebra, and matrix algebra of this model in the next three sections. In the fourth section, we discuss a number of WMDS details. Then we present an example. Finally, we discuss two statistics developed for WMDS and flattened weights.

Geometry of the Weighted Euclidean Model

The weighted Euclidean model assumes that the individuals vary in the importance they attach to the dimensions of the stimulus space *X*. While one individual may perceive one of the dimensions as being more important than another, another individual may have just the opposite perception. The notion of salience is incorporated into the model by weights w_{ka} for each individual *k* on each dimension *a*. These weights vary from 0.0 to 1.0. If the weight is large (near 1.0), then the dimension is relatively important; if it is small (near 0), the dimension is not so important.

Geometrically, the weighted Euclidean model represents individual differences in a special space called the **weight space**. In this space, individuals are represented by vectors emanating from the origin of the space. The direction of the vector from the origin represents the relative weighting of each dimension. The length of the vector represents the overall salience of the dimension to the individual.

A schematic diagram of the geometry of the weighted Euclidean model is presented in Figure 14-28. At the top are two hypothetical data matrices concerning four stimuli (potato, spinach, lettuce, and tuna). These matrices are labeled S_2 and S_5, indicating that they are matrices number 2 and 5 of (we assume for this example) five such matrices. (Note that the MDS algorithm would not actually analyze data matrices that have only four rows and columns, since they are too small to support meaningful results.)

Figure 14-28

Geometry of the weighted Euclidean model (not drawn in SPSS)

Data Matrices

S_2

	Potato	Spinach	Lettuce	Tuna
Potato		4	2	3
Spinach	4		1	6
Lettuce	2	1		5
Tuna	3	6	5	

S_5

	Potato	Spinach	Lettuce	Tuna
Potato		1	3	6
Spinach	1		2	5
Lettuce	3	2		4
Tuna	6	5	4	

Group Coordinates X

	I	II
x_1	-2	1
x_2	-1	4
x_3	1	3
x_4	4	-1

Matrix Weights W

	I	II	Length
w_1	0	0.9	0.90
w_2	0.2	0.8	0.82
w_3	0.6	0.6	0.85
w_4	0.4	0.4	0.56
w_5	0.8	0.2	0.82

Group Distances D

	Potato	Spinach	Lettuce	Tuna
Potato		3.16	3.61	6.32
Spinach	3.16		2.24	7.07
Lettuce	3.61	2.24		5.00
Tuna	6.32	7.07	5.00	

In the middle of Figure 14-28 are two spaces, each two-dimensional. The space on the left is the hypothetical group stimulus space X. In it are four points labeled x_1 through x_4. These points correspond to the four stimuli. The space on the right is a hypothetical weight space W with five vectors labeled w_1 through w_5. These vectors correspond to the five data matrices. Note that the dimensionality of X is always the same as that of W in WMDS.

At the bottom of Figure 14-28, we present numerical information that corresponds to the diagrams in the middle of the figure; on the left is the group-stimulus-coordinates matrix X and in the middle the weights matrix W. The two columns of the matrix X correspond to the two dimensions of the stimulus space. The rows of this matrix specify the positions of each point on each dimension of the stimulus space. The first two columns of the weight matrix W correspond to the two dimensions of the weight space. The values in these columns for each row specify the location of the tips of the weight vectors in the weight space. The third column contains the length of each weight vector in the weight space, which is the square root of the sum of the squares of the other values in the row of the weight matrix. At the bottom-right of the figure is the matrix D of Euclidean distances between the points in the group stimulus space X.

The stimulus space X is the same as that in RMDS and CMDS. For this reason, we use the same notation as used previously. We call the space the **group stimulus space** for the same reasons we called it that for RMDS. The group Euclidean distances D are also the same as the Euclidean distances in the (unweighted) Euclidean model, except that they are for the group stimulus space, and so are called **group distances**. The group space and its distances present information about how the entire group of individuals (or whatever group of things generated the several data matrices) structures the stimuli and carries the same interpretation as that for RMDS, with one major (and one minor) exception to be discussed next. Note, however, that this group information does not represent the structure for any individual, as each individual's structure is modified from the group structure by the individual's weights w_k. Rather, the group space represents the information that is shared in common by all the individuals about the structure of the stimuli.

The stimulus space X has the same characteristics for WMDS as it does for CMDS and RMDS, except for two differences. One very important difference is that for WMDS, the stimulus space is not rotatable, as will be proven in the matrix algebra section below. The other difference, which is less important but occasionally confusing, is that the dimensions can be stretched or shrunk by separate scaling factors, whereas the Euclidean model requires that the same scaling factor be used for all dimensions. The two models are the same in that for both models, the dimensions can be reflected, permuted, and translated.

The fact that the stimulus space is not rotatable is very important because it implies that the dimensions themselves should be meaningful. We should, if the model describes the data accurately, be able to directly interpret the dimensions of the space, not having to worry about rotations of the dimensions into another orientation for interpretation. It should be kept in mind, however, that the stability of the dimensions depends on two things: the goodness of fit of the analysis and the variation in the weights. First, if there is no variation in the orientation of the weight vectors, the dimensions can be rotated. Also, if there is little variation in the weight vectors' orientations, the orientation of the dimensions is not as tightly determined as when there is great variation. Second, if the model fits the data perfectly, the dimensions are completely stable (unless there is no variation in the weights). However, the degree of stability decreases as the fit decreases. Thus, for a poorly fitting analysis, the orientation of the dimensions is not as well determined as for an analysis that fits very well.

Now we turn to the weights W. Each individual (each matrix of data) is represented by a weight vector, the vectors being labeled w_1 through w_5 for the five individual matrices. Notice that the weight space has vectors only in the positive quadrant. Generally, only positive weights are interpretable. Thus, by default SPSS restricts weights to be non-negative. Notice also that no weights are greater than 1.00. This is because they have been normalized so that their length equals the proportion of variance accounted for in the individual's data by the model (the RSQ value discussed earlier).

In the weighted Euclidean model, individual differences in perception or cognition are represented by differences in the orientation and length of the vectors w_k in the weight space W. Of these two aspects of a vector, variation in orientation of the vectors is most important, since it reflects differences in the importance of the dimensions to the individuals. Two vectors that point in exactly the same direction indicate that the two individuals have the same relative weighting of the dimensions, regardless of the length of the vectors. The different lengths simply indicate that one person's data are better described by the analysis than the other's, the longer vector (and larger weights) representing the person whose data are better fit.

The nature of the individual differences can be seen most readily by comparing the personal stimulus spaces for the several individuals. The personal stimulus space for an individual is what results after applying the (square root of the) weights for an individual to the group space. The weights shrink the dimensions of the group space, with important dimensions being shrunk less than unimportant dimensions (since weights near 1.00 represent important dimensions and will shrink a dimension relatively little). Thus, in the personal space, important dimensions are longer than unimportant dimensions. The algebra for this will be shown in the next section.

The idea of an individual's personal space is illustrated in Figure 14-29 for three individuals. Information at the top of the figure is for individual 2, at the middle for individual 5, and at the bottom for individual 4.

Figure 14-29
Personal space structures (not drawn in SPSS)

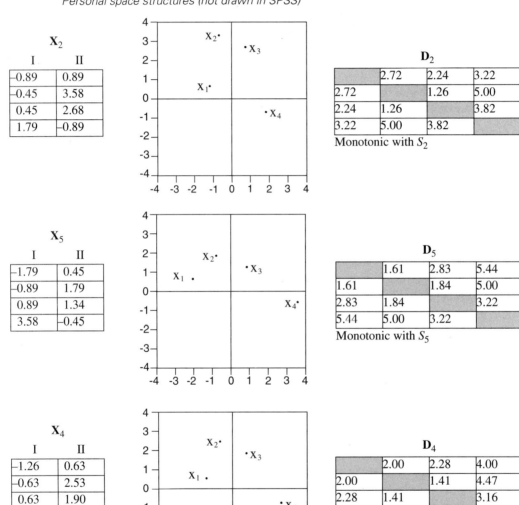

X₂

I	II
-0.89	0.89
-0.45	3.58
0.45	2.68
1.79	-0.89

D₂

	2.72	2.24	3.22
2.72		1.26	5.00
2.24	1.26		3.82
3.22	5.00	3.82	

Monotonic with S_2

X₅

I	II
-1.79	0.45
-0.89	1.79
0.89	1.34
3.58	-0.45

D₅

	1.61	2.83	5.44
1.61		1.84	5.00
2.83	1.84		3.22
5.44	5.00	3.22	

Monotonic with S_5

X₄

I	II
-1.26	0.63
-0.63	2.53
0.63	1.90
2.53	-0.63

D₄

	2.00	2.28	4.00
2.00		1.41	4.47
2.28	1.41		3.16
4.00	4.47	3.16	

Monotonic with S_4

Consider the information for individual 2 at the top of the figure. At the top left is this individual's matrix of personal coordinates, labeled X_2. (Note that X_2 is a matrix of coordinates for all the stimuli in the personal space for individual 2, whereas x_2 is a row of coordinates for stimulus 2 in the group stimulus space.) At the top middle is the personal space X_2 for this individual (the coordinates are given in the matrix X_2). At the top right is the individual's matrix of personal distances, labeled D_2, which are unweighted Euclidean distances between the points in the personal space X_2, and weighted Euclidean distances between the points in the group space X as weighted by the square root of the weights w_2 for this individual.

It is informative to compare the personal space structures for each of these three individuals with their weights. Individual 2 has weights of 0.2 and 0.8, as can be seen at the bottom middle of Figure 14-28. Thus, this individual finds dimension 2 four times as important as dimension 1. This is reflected in the top-middle of Figure 14-29 by the fact that dimension 2 is relatively longer than dimension 1. For this person, the personal space is mostly the vertical dimension.

Individual 5 has weights that are opposite to individual 2's—0.8 and 0.2. These weights indicate that individual 5 finds dimension 2 to be one-fourth as important as dimension 1. We see, in the middle of Figure 14-29, that person 5's personal space is much longer along dimension 1 than dimension 2 (the space is mostly horizontal).

Individual 4, on the other hand, finds the two dimensions to be equally important, having weights of 0.4 and 0.4 on both dimensions. Thus, this person's own stimulus structure is the same as the group's, except that it has shrunk due to the relatively small weights.

The last point we want to make about the geometry of WMDS is that one individual's personal distances are not related to another individual's personal distances by any type of simple one-to-one function. In particular, they are not linearly related, nor are they even monotonically related. As can be seen in Figure 14-29, person 2's distances D_2 are not monotonic with the distances for person 5 or person 4. This implies that it is possible to use the weighted Euclidean model to perfectly describe data in several dissimilarity data matrices even though the data are not even monotonically related to each other. In fact, the top two distance matrices in Figure 14-29 (D_2 and D_5) are monotonically related to the two data matrices S_2 and S_5 in Figure 14-28, even though the data matrices are not monotonically related to each other. Thus, the structure in X and W perfectly describes the data from these two people, even though their data are not simply related to each other. This is an important distinction from RMDS, where the model assumes that all data matrices are monotonically (or linearly) related to each other, except for error.

Algebra of the Weighted Euclidean Model

WMDS is based on the following definition of weighted Euclidean distance:

$$d_{ijk} = \left[\sum_a^r w_{ka}(x_{ia} - x_{ja})^2\right]^{\frac{1}{2}}$$

where $0 \le w_{ka} \le 1$, $r \ge 2$, and d_{ijk} is the distance between stimuli i and j as perceived by subject k. We discuss the algebraic characteristics of the weighted distance model in the next several sections, first presenting several concepts in scalar algebra, then several more concepts in matrix algebra.

Group Stimulus Space

The x_{ia} are the same stimulus coordinates as those in the CMDS and RMDS situations, with the new restriction that one-dimensional solutions are not permitted. Taken together, the x_{ia} form the $n \times r$ stimulus coordinates matrix X. A row of this matrix is denoted x_i and contains all r of the coordinates for stimulus i. As in the RMDS case, the stimulus space X represents the structure of the stimuli as perceived by the entire group of individuals. Thus, it is called the group stimulus space.

As mentioned above, the WMDS group stimulus space has the same characteristics as the RMDS group stimulus space, except that it is not rotatable and the dimensions can be arbitrarily rescaled by separate constant factors. We discuss the rotation issue in the matrix algebra section that follows. Here we discuss the rescaling issue.

In WMDS, we can rescale each dimension a by a unique constant c_a if we rescale the dimensions of the weight space by corresponding constants c_a^{-2}. This can be shown by noting that

$$d_{ijk} = \left[\sum_a^r w_{ka}c_a^{-2}(c_a x_{ia} - c_a x_{ja})^2\right]^{\frac{1}{2}}$$

$$d_{ijk} = \left[\sum_a^r w_{ka}c_a^{-2}c_a^2(x_{ia} - x_{ja})^2\right]^{\frac{1}{2}}$$

$$d_{ijk} = \left[\sum_a^r w_{ka}(x_{ia} - x_{ja})^2\right]^{\frac{1}{2}}$$

which is the basic WMDS equation. This implies that there is an arbitrary normalization of the dimensions that must be defined. This normalization and its implications are discussed below. Note that this arbitrary aspect does not exist in the Euclidean model, since there are no weights that can be rescaled in a way that compensates for rescalings of the stimulus space.

The w_{ka} in the equation above represent the weights that individual k associates with the dimensions a of the stimulus space. As noted above, the weights are collected together into an $m \times r$ matrix W, which has one row for each of the m individuals (data matrices) and one column for each of the r dimensions.

For WMDS, the configuration derived by SPSS is normalized as follows. For all models, the coordinates are centered:

$$\sum_{i}^{n} x_{ia} = 0$$

and the length of the stimulus space dimensions are each set equal to the number of points:

$$\sum_{i}^{n} x_{ia}^{2} = n$$

The weights are normalized so that

$$\sum_{a}^{r} w^{2}_{ka} = r_{k}^{2}$$

where r_{k}^{2} is RSQ, the squared correlation between subject k's squared weighted Euclidean distances D_{k}^{2} and the same subject's dissimilarity data S_{k}. This normalization has the characteristic that the sum of an individual's squared weights reflects the proportion of the total variance of the individual's transformed data S_{k} that is accounted for by the model.

Note that there is a subtle difference in normalization of the stimulus space in the weighted and unweighted models. In the weighted model, the dimensions of the stimulus space X are each separately normalized to be of equal length n, while in the unweighted model, the dimensions are jointly normalized to have an average length of n. Thus, in CMDS and RMDS, you interpret the relative length of the stimulus space dimensions as indicating their relative importance, but in WMDS, you cannot interpret the relative importance of stimulus space dimensions because they are arbitrarily normalized to be equal. The relative importance of WMDS dimensions is found in the average weights on a dimension, rather than in the spread of the stimulus coordinates. This will be seen in the example below.

Personal Spaces

One essential difference between the weighted and unweighted Euclidean models is that in WMDS, the coordinates x_{ia} of the group stimulus space X can be weighted by the square root of an individual's weights w_{ka} to obtain the individual's personal stimulus space. The coordinates x_{ika} of stimuli in an individual's personal stimulus space X_k are derived from the group space X and the weights W by the equation:

$$x_{ika} = x_{ia}(w_{ka})^{\frac{1}{2}}$$

It can be shown that the distances d_{ijk} for an individual can be reexpressed in terms of the coordinates x_{ika} as the simple Euclidean distances in the personal stimulus space:

$$d_{ijk} = \left[\sum_{a}^{r} (x_{ika} - x_{jka})^2 \right]^{\frac{1}{2}}$$

Matrix Algebra of the Weighted Euclidean Model

To state the weighted Euclidean model in matrix algebra, we must first define a new set of weight matrices W_k. There are m of these weight matrices W_k, one for each of the m individuals. Each W_k is an $r \times r$ diagonal matrix, with the weights for individual k on the diagonal.

Note that the new diagonal weight matrices W_k are not the same as the previously defined rectangular weight matrix W. However, the new diagonal matrices contain the same information as the earlier rectangular matrix; the rows w_k of the earlier W have become the diagonals of the new W_k (note that lowercase w_k is a row of W, whereas uppercase W_k is a diagonal matrix). Furthermore, the diagonal elements w_{kaa} of matrix W_k correspond to the elements w_{ka} of the kth row of the earlier matrix W.

Now that we have introduced the new notation W_k, we can state the weighted Euclidean model in matrix algebra terminology. The model is

$$d_{ijk} = [(x_i - x_j)W_k(x_i - x_j)']^{\frac{1}{2}}$$

Personal Spaces

We mentioned above that an essential aspect of the weighted model is the notion of an individual's personal stimulus space. The personal space for individual k is represented by the matrix $\mathbf{X}_k$, which is defined as

$$\mathbf{X}_k = \mathbf{X}\mathbf{W}_k^{\frac{1}{2}}$$

giving an alternative expression for the weighted Euclidean model as

$$d_{ijk} = [(x_{ik} - x_{jk})(x_{ik} - x_{jk})']^{\frac{1}{2}}$$

where x_{ik} is the ith row of $\mathbf{X}_k$. We see then that this is the same as the formula for Euclidean distance in individual k's personal space.

Rotation

It is very important to note that the dimensions of the WMDS joint space cannot be orthogonally rotated without violating the basic definition of the model (unlike the CMDS and RMDS dimensions, which can be rotated). This can be seen by defining the rotated stimulus space:

$$\mathbf{X}^* = \mathbf{X}\mathbf{T}$$

where T is an $r \times r$ orthogonal rotation matrix such that $\mathbf{TT'} = \mathbf{T'T} = \mathbf{I}$. It follows that the distances between points in the rotated space are

$$d_{ijk} = [(x_{ik}^* - x_{jk}^*)\mathbf{W}_k(x_{ik}^* - x_{jk}^*)']^{\frac{1}{2}}$$

$$d_{ijk} = [(x_{ik} - x_{jk})\mathbf{T}\mathbf{W}_k\mathbf{T}'(x_{ik} - x_{jk})']^{\frac{1}{2}}$$

$$d_{ijk} = [(x_{ik} - x_{jk})\mathbf{W}_k^*(x_{ik} - x_{jk})']^{\frac{1}{2}}$$

While it appears that the distances defined by the last equation satisfy the definition of the weighted Euclidean model, the matrix $\mathbf{W}_k^*$ in the last equation is not diagonal, thus violating the definition of the model. Therefore, orthogonal rotation is not allowed.

Details of WMDS

In this section, we discuss the distinction between metric and nonmetric WMDS, the measures of fit in WMDS, and the fundamental WMDS equation.

Metric and Nonmetric WMDS

WMDS is appropriate for the same type of data as RMDS. However, RMDS generates a single distance matrix D, while WMDS generates m unique distance matrices D_k, one for each data matrix S_k. Just as in RMDS, the data can be symmetric or asymmetric, matrix or row conditional, have missing values, and be at the ordinal, interval, or ratio levels of measurement. Thus, we can have metric or nonmetric WMDS, depending on the measurement level.

The distances D_k are calculated so that they are all as much like their corresponding data matrices S_k as possible. For metric WMDS (when the data are quantitative), the least-squares problem is

$$l_k\{S_k\} = D_k^2 + E_k$$

and for nonmetric WMDS (qualitative data), the least-squares problem is

$$m_k\{S_k\} = D_k^2 + E_k$$

Thus, for WMDS, SPSS solves for the $n \times r$ matrix of coordinates X, for the m diagonal $r \times r$ matrices W_k, and also for the m transformations m_k or l_k. It does this so that the sum of squared elements in all error matrices E_k is minimal when summed over all matrices and when subject to normalization constraints on X and W_k. The scaling algorithm used by SPSS (ALSCAL) was the first to incorporate both nonmetric and metric WMDS, as well as the other types of MDS discussed above, and is considered to be the third breakthrough in multidimensional scaling (Takane et al., 1977).

Measures of Fit

The measures of fit (S-stress, stress, and RSQ) are all defined in the same way as RMDS, except that the weighted Euclidean distances D_k are used in the measures instead of the unweighted distances.

Fundamental WMDS Equation

We can now summarize, in a single equation, the WMDS data analysis problem being solved by the MDS procedure. The fundamental equation is

$$S_k \overset{L}{=} T_k = D_k^2 + E_k$$

This equation parallels the fundamental CMDS and RMDS equations and is to be understood in a similar fashion.

Example: Perceived Body-Part Structure

We return to the Jacobowitz data (Young, 1974) concerning how children and adults understand the relationship between various parts of the body. The data are the same as those discussed for RMDS (see Table 14-1).

To obtain a WMDS solution, open the data file *combinedbody.sav* and from the menus choose:

Analyze
  Scale
    Multidimensional Scaling (ALSCAL)...

     ▶ Variables: cheek, face, mouth, head, ear, body, arm, elbow, hand, palm, finger, leg, knee, foot, toe
     Distances
     ⊙ Data are distances
     Shape: Square asymmetric

Model...
    Level of Measurement
    ⊙ Ordinal
    Conditionality
    ⊙ Row
    Dimensions
    Minimum: 3 Maximum: 3
    Scaling Model
    ⊙ Individual differences Euclidean distance

Options...
    Display:
    ☑ Group plots
    ☑ Data matrix
    Criteria
Minimum s-stress value: .001

The WMDS results are shown in Figure 14-30 and Figure 14-31. These are the results of applying the WMDS model to the data for both the children and adults simultaneously. Figure 14-30 displays a three-dimensional stimulus space X (with the origin, which is shown, at the center), and Figure 14-31 shows the positive portion of the three-dimensional weight space W (the origin is where the sides all intersect).

We can see in Figure 14-30 that the stimulus space displays the overall hierarchical structure that was hypothesized by Jacobowitz to exist in this data. We can also see in Figure 14-31 that the weights for adults and children occupy different portions of the space, implying that adults and children have different perceptions/cognitions concerning the body parts.

Figure 14-30
Three-dimensional stimulus space (not drawn in SPSS)

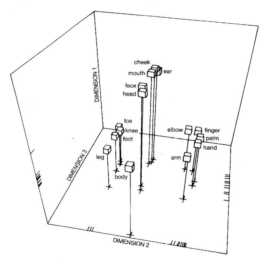

Figure 14-31
Positive portion of the three-dimensional weight space (not drawn in SPSS)

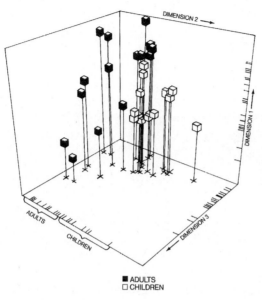

■ ADULTS
□ CHILDREN

The weight space shows that the adults generally have a relatively small weight on dimension 2 (they are all on the left side of the space), whereas the children generally have a relatively small weight on dimension 3 (they are at the back of the space). This gives us a way to look at individual differences between adults and children. We can construct a hypothetical adult who has no weight on dimension 2 and a hypothetical child who has no weight on dimension 3 and investigate what structures such people would have.

A hypothetical child who has no weight on dimension 3 is one who collapses the three-dimensional space into a two-dimensional space consisting of only dimensions 1 and 2. Such a space is presented in Figure 14-32. We see that this space presents a very simple version of the hierarchical structure, showing only that this hypothetical child differentiates "body" from the other body parts, which in turn are clustered according to whether they are arm parts, leg parts, or head parts. This is a structure for a very young child (in fact, it is rather like the structure derived from the data of the youngest age group, which we do not analyze in this chapter).

Figure 14-32
Hypothetical child—no weight for dimension 3 (not drawn in SPSS)

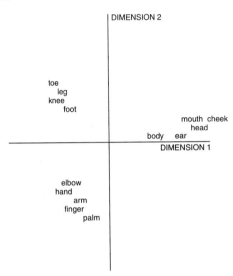

A hypothetical adult who has no weight on dimension 2 is one who collapses the three-dimensional space into a two-dimensional space consisting of only dimensions 1 and 3. This space is presented in Figure 14-33. We see that this space displays both the hierarchy and the classification. Thus, this space represents the structure of a hypothetical individual who has the most developed understanding of the relationships among body parts.

Figure 14-33
Hypothetical adult—no weight for dimension 2 (not drawn in SPSS)

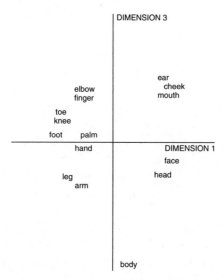

The results shown in Figure 14-30 through Figure 14-33 are based on a three-dimensional WMDS analysis. The data are the same as those shown for the RMDS analysis. We have specified a three-dimensional solution because the RMDS analyses suggested that a three-dimensional solution is appropriate. We have also requested the weighted Euclidean (INDSCAL) model.

Figure 14-34 shows the iteration history for the three-dimensional solution. The initial iteration has an S-stress value of 0.29931. After seven iterations, the value has improved modestly to 0.26620. At this point, iterations stop because the S-stress improvement is less than 0.001. As with the RMDS analysis, the fit is good for this amount of data, but it is not excellent.

Figure 14-34
Iteration history for the three-dimensional solution

```
Iteration history for the 3 dimensional solution (in squared distances)

            Young's S-stress formula 1 is used.

        Iteration     S-stress      Improvement
            0           .29931
            1           .29705
            2           .28085        .01620
            3           .27488        .00596
            4           .27188        .00300
            5           .27003        .00185
            6           .26874        .00129
            7           .26777        .00097

                  Iterations stopped because
          S-stress improvement is less than    .001000
```

SPSS then generates 450 stress and RSQ indexes (not shown), one for each of the 15 rows of the 30 matrices of data. In addition, there are 30 stress and RSQ indexes for each matrix of data. Then the RSQ (but not the stress) is calculated for each stimulus, averaged over matrices (this appears only for asymmetric, row conditional data). Finally, the RSQ and stress indexes averaged over all of the data are calculated. This last stress measure is 0.222, which does not equal the S-stress of 0.26620 because these formulas are defined differently, as discussed above.

The RSQ values for each matrix (not shown) vary between 0.398 and 0.969, indicating that there are no judges who are being fit very poorly, which in turn indicates that none of the judges are giving purely random judgments. There is some suggestion that the adults fit better than the children, as the adult RSQs are higher (ranging from 0.604 to 0.969) than the children's (0.398 to 0.936).

The RSQ values for each stimulus vary between 0.704 and 0.810, except for the term "body," which fits noticeably better at 0.868. There seems to be a bit less error, or a bit more agreement, in the placement of "body" in the overall structure. There are no extremely low RSQs for any given row of any given matrix (suggesting that none of the judged rank orders are reversed, a not uncommon problem).

Figure 14-35 shows the stimulus coordinates (*X*), and Figure 14-36 shows the subject weights (*W*).

Figure 14-35
Stimulus coordinates

Dimension

Stimulus Number	Stimulus Name	1	2	3
1	cheek	1.3626	.4341	.8810
2	face	1.3837	.5130	-.4782
3	mouth	1.2937	.5572	.8332
4	head	1.3274	.4390	-.7622
5	ear	1.2873	.4263	1.0756
6	body	.1828	.1306	-2.8441
7	arm	-.5068	-1.2064	-.9688
8	elbow	-.6171	-1.1553	.6877
9	hand	-.4572	-1.4549	-.0229
10	palm	-.3129	-1.5981	.3821
11	finger	-.5307	-1.3534	.7875
12	leg	-1.0278	1.0708	-.7948
13	knee	-1.1487	1.0835	.3344
14	foot	-1.1116	1.0605	.1649
15	toe	-1.1246	1.0531	.7247

Figure 14-36
Subject weights and weirdness index

```
Subject weights measure the importance of each dimension to each subject.
Squared weights sum to RSQ.

A subject with weights proportional to the average weights has a weirdness of
zero, the minimum value.
A subject with one large weight and many low weights has a weirdness near one.
A subject with exactly one positive weight has a weirdness of one,
the maximum value for nonnegative weights.
```

Subject Weights

Subject Number	Weird- ness	Dimension 1	2	3
1	.2723	.7010	.5758	.2109
2	.2305	.4320	.4568	.2206
3	.1693	.4460	.4177	.3114
4	.3341	.6441	.6722	.2208
5	.2281	.4977	.5515	.2903
6	.1493	.7487	.5132	.2651
7	.2020	.3947	.4223	.2566
8	.1990	.4487	.4524	.2366
9	.3247	.6780	.6681	.2194
10	.1814	.4326	.4388	.2791
11	.3661	.6361	.6961	.2100
12	.1330	.5225	.3696	.3338
13	.4693	.8585	.3704	.1025
14	.1503	.5041	.4608	.3293
15	.3859	.5751	.5913	.1592
16	.2659	.8187	.4254	.2000
17	.7846	.3842	.0928	.8281
18	.1971	.7665	.4876	.2320
19	.4360	.4317	.3532	.5427
20	.2459	.7664	.3991	.1984
21	.4258	.7679	.4145	.1050
22	.4725	.9041	.1454	.2855
23	.4150	.8484	.4743	.1219
24	.4116	.9019	.1958	.2758
25	.7125	.9105	.3052	.0000
26	.0436	.7525	.5286	.3490
27	.6492	.2883	.2988	.7114
28	.2313	.7801	.3310	.4265
29	.5126	.6725	.2157	.6322
30	.4585	.7018	.2432	.5963
Overall importance of each dimension:		.4418	.1974	.1268

Displayed with the subject weights is the weirdness index, whose interpretation is briefly noted on the output and is explained extensively in the next section. It is useful to look over the values of the weirdness index for values near 1.0, as these values indicate that the associated subject's weights w_k are unusual. We see that some of the weirdness values for the children (first 15 values) range up to 0.48, whereas these values range up to 0.79 for adults (subjects 16 through 30), since there are five adults with values higher than the children's values. This suggests that there are more adults than children who have at least one very low weight, indicating that more of the adults tend to have nearly two-dimensional solutions.

Figure 14-37 shows an SPSS plot of the stimulus coordinates used to create the display in Figure 14-30.

Figure 14-37
Plot of stimulus coordinates

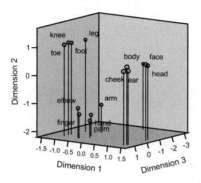

Figure 14-38, Figure 14-39, and Figure 14-40 show the plots of derived subject weights. To obtain the two-dimensional plots from the three-dimensional subject weight plot, open the three-dimensional plot in the Chart Editor by double-clicking on it; from the Edit menu choose Properties. Then exclude each of the variables as necessary by moving them to the Excluded list. Arrange the remaining variables on the *x* and *y* axis.

Individuals are identified in these plots by the subject numbers shown in Figure 14-36. The correct interpretation of the three plots shown here is that the points plotted for each individual represent the ends of vectors drawn from the origin of the space. We interpret both the direction and the length of each vector. If two vectors point in the same direction, they represent subjects who have equivalent weighting schemes (the ratios of one subject's weights on the various dimensions are the same as the ratios of the other subject's weights) and, therefore, the same stimulus structure in their personal stimulus spaces. This is true even if the vectors are of different lengths, since the length simply indicates goodness of fit (the squared length of a vector—that is, the sum of the squared weights—equals RSQ).

Figure 14-38
Plot of derived subject weights, dimension 1 versus dimension 2

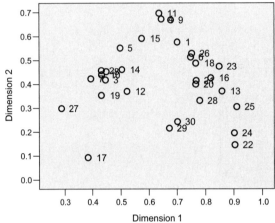

Figure 14-39
Plot of derived subject weights, dimension 1 versus dimension 3

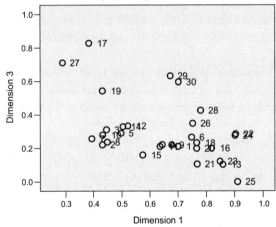

Figure 14-40
Plot of derived subject weights, dimension 2 versus dimension 3

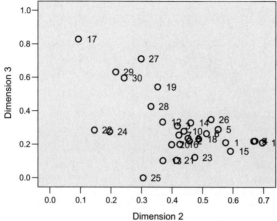

With this in mind, let's look at the three weight space plots. In Figure 14-38, we note that if we draw a line from the origin through the subject whose plot symbol is 12 (in the middle of the plot), all the subjects below the line are adults (with two exceptions) and all the subjects above the line are children (with four exceptions). Thus, essentially all adults weight dimension 1 (horizontal axis) more heavily relative to dimension 2 (vertical axis) than is the case for children. More simply stated, adults find dimension 1 more salient than children do.

In Figure 14-39, which displays dimensions 1 and 3, we can draw a similar line from the origin to just below the subject whose plot symbol is 30. The subjects above this line are adults. Thus, some of the adults, but none of the children, find dimension 1 more important than dimension 3. Furthermore, the adults who are below this line are all further from the origin than the children, suggesting that children find neither dimension 1 nor dimension 3 as salient as adults do.

Looking at Figure 14-40, which displays dimensions 2 and 3, we conclude that some adults find dimension 3 more salient than 2, whereas all children find dimension 2 more salient than 3. Putting all of these observations together, we arrive at the interpretation presented in the results section above.

Figure 14-41 shows the scatterplot of linear fit, which presents the plot of all of the $15 \times 14 = 210$ transformed dissimilarities for each of the 30 matrices T_k (6,300 elements in total) versus their corresponding elements in the matrices D. We see that this plot shows fairly little scatter, particularly when one considers all the points along the diagonal of the plot. However, there is a definite suggestion of greater scatter for the small distances than for the large distances, an artifact explained above.

Chapter 14

Figure 14-41
Scatterplot of linear fit

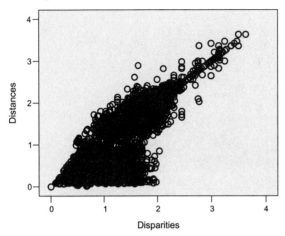

The matrix and plot of flattened subject weights appear in Figure 14-42 and Figure 14-43. In Figure 14-43, individuals are identified by the subject numbers shown in Figure 14-42. The meaning of the flattened weights, and their interpretation, is discussed below.

Figure 14-42
Flattened subject weights

```
Flattened Subject Weights

                          Variable
Subject    Plot       1          2
Number     Symbol
   1          1      .0333       .7634
   2          2     -.6598      1.0028
   3          3     -.7429       .4575
   4          4     -.4085      1.2504
   5          5     -.8108      1.0027
   6          6      .1948       .2700
   7          7     -.8440       .8247
   8          8     -.6175       .8660
   9          9     -.2896      1.1480
  10          A     -.7729       .7089
  11          B     -.4644      1.3865
  12          C     -.3477      -.0654
  13          D     1.5036      -.2908
  14          E     -.6585       .4632
  15          F     -.2834      1.3352
  16          G      .8440      -.1317
  17          H    -1.4647     -2.2960
  18          I      .4106       .1928
  19          J    -1.2038      -.4085
  20          K      .8015      -.1508
  21          L     1.0943       .1330
  22          M     1.7784     -1.9303
  23          N     1.0168       .1948
  24          O     1.6041     -1.6044
  25          P     2.3856      -.5533
  26          Q     -.0478       .1555
  27          R    -2.0773      -.7565
  28          S      .3393      -.9002
  29          T     -.2115     -1.6108
```

Figure 14-43
Plot of flattened subject weights

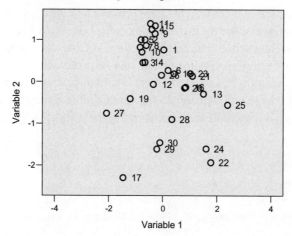

Weirdness Index

It has been pointed out by Takane et al. (1977) and especially by MacCallum (1977) that the weights used to represent individual differences in the INDSCAL model are commonly misinterpreted. The correct interpretation of the weights is that they represent the end of a vector directed from the origin of the weight space. Thus, the angle of each subject's vector, relative to the dimensions of the space (and relative to other subjects' vectors), is interpretable. A common mistake is to treat the weights as a point and interpret the distances between points, when instead the weights should be treated as a vector and the angles between vectors should be interpreted. Thus, a common misinterpretation is that subject weight points that are "near each other" in the subject weight space are "similar," when in fact the correct interpretation is that subject weight vectors which are oriented in "roughly the same direction" are "similar."

The **weirdness index** is designed to help interpret subject weights. The index indicates how unusual or weird each subject's weights are relative to the weights of the typical subject being analyzed. The index varies from 0.00 to 1.00. A subject with a weirdness of 0.00 has weights that are proportional to the average subject's weights (to the mean weights). Such a subject is a totally typical subject. As the weight ratios become more and more extreme, the weirdness index approaches 1.00. Finally, when a subject has only one positive weight and all the remaining weights are 0, the weirdness index is 1.00. Such a subject is very weird, using only one of the dimensions of the analysis.

Consider the hypothetical subject weight space presented in Figure 14-44. In this figure, we present the weight vector for just one subject. It is shown as a vector directed from the origin of the weight space. The endpoint of the vector, when projected onto dimensions 1 and 2, has Cartesian coordinates w_{k1} and w_{k2}, respectively. In the usual interpretation of individual differences, the subject is said to have weight w_{k1} on dimension 1 and weight w_{k2} on dimension 2, the weights representing the degree of importance (salience or relevance) of the dimensions to the subject.

Figure 14-44
Hypothetical subject weight space (not drawn in SPSS)

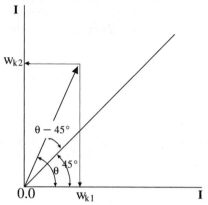

The Cartesian coordinates can be converted into polar coordinates without loss of information. The polar coordinates of vector w_k are the length of the vector and its angle relative to dimension 1. We can represent the length as r_k (not labeled in Figure 14-44) and the angle as θ_k (labeled as θ in the figure). The length r_k is set equal to the square root of the RSQ value displayed for each subject. Thus, the usual representation of individual differences, namely w_{k1} and w_{k2}, can be changed to the wholly equivalent representation r_k and θ_k. When there are more than two dimensions, we still have r_k but several θ_{ka}, one for each dimension a except the last.

The reference axis for the angle is arbitrary, of course. We could have chosen dimension 2 instead of dimension 1. In fact, we could have chosen any other direction through the subject space. The $45°$ line through Figure 14-44, for example, could serve as the reference for a new angle $\theta - 45°$, and we could express the location of the subject's vector by its length and the angle between it and the $45°$ line.

Consider the $45°$ line. It has the simple interpretation that it represents equal weighting of both dimensions. Furthermore, as a subject's weight vector departs from this line, the subject has a more and more extreme pattern of weights. In the extreme, if the subject's vector lies along one of the dimensions, the subject is using that dimension to the complete exclusion of the other dimension.

This interpretation of the $45°$ line is, however, entirely dependent on a particular way of arbitrarily normalizing the subject weights. As is shown in the left portion of Figure 14-45, SPSS normalizes the subject weights so that on average dimension 1 gets the heaviest weighting, dimension 2 the next heaviest, and so forth. As mentioned above, this normalization is arbitrary. If instead we were to normalize so that the $45°$ line represented the typical subject's weights (instead of equal weights), the line would correspond to the average subject. This modification, which is shown in the right portion of Figure 14-45, is done by simply defining a new set of weights:

$$w'_{ka} = w_{ka} / \left(\sum_{k}^{n} w_{ka} \right)$$

When there are more than two dimensions, this normalization orients the typical subject's weight vector along the line that is at a $45°$ angle to all dimensions.

Figure 14-45
Normalized subject weights (not drawn in SPSS)

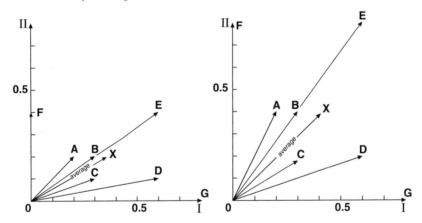

If we make this modification in the normalization of the weights, departures from the $45°$ line have a very nice interpretation. First, a subject whose vector is on the $45°$ line is one who weights the dimensions just as the typical subject does. Second, as a subject's vector departs farther and farther from the line, the subject becomes less and less like the typical subject.

While this normalization has desirable aspects, it is used only in the definition of the weirdness index. In particular, SPSS does not normalize the weights in this fashion because it seems to be more desirable to continue with the normalization conventions adopted in other MDS programs. However, this normalization is involved in calculating the weirdness index.

The final steps in defining a convenient index of a subject's "weirdness" (that is, how far the subject's weight vector is from the average subject's vector) are first to determine the angle between the subject's normalized weight vector and the 45° line, and second to define a function of this angle that varies from 0.00 to 1.00. We do this by defining standardized vectors v_k that are unit-length and collinear with the weight vectors w_k:

$$v_{ka} = w'_{ka} / \left[\sum_a^r (w'_{ka})^2 \right]^{\frac{1}{2}}$$

The angle between a subject's weight vector and the 45° line can now be defined, in radians, as

$$\cos^{-1}\left[r^{-\frac{1}{2}} \left(\sum_a^r v_{ka} \right) \right]$$

This angle varies from a minimum that occurs when the subject's vector coincides with the 45° line (that is, when the subject's weights are proportional to the average weights) to a maximum value that occurs when the subject's vector coincides with one of the edges of the subject's weight space (that is, when the subject has only one positive weight). A little investigation shows that the maximum angle, in radians, is $\cos^{-1} r^{-1/2}$. Thus, if we divide by this maximum value, we obtain an index that varies from 0.00 to 1.00. Finally, we divide the previous formula by this maximum value to obtain the complete formula for the weirdness index:

$$\left(\cos^{-1}\left[r^{-\frac{1}{2}} \left(\sum_a^r v_{ka} \right) \right] \right) / \left(\cos^{-1} r^{-\frac{1}{2}} \right)$$

Flattened Weights

As pointed out above, the subject weights in WMDS are often misinterpreted. A common mistake is to interpret the distance between subject points instead of interpreting the angle between subject vectors. In addition, standard statistical procedures are inappropriate for interpreting these weights because the weights present angular information, not linear information.

Because of these characteristics of the weights, SPSS calculates and displays information called **flattened weights**. The weight vectors, when flattened, become weight points. The angles between the vectors, when flattened, become distances between the points. Thus, the flattened weight variables are ordinary linear variables that can be used in ordinary statistical procedures. However, because of lack of independence between the weights, hypothesis testing is inappropriate. The statistical procedure should be used in a descriptive fashion only.

Flattened weights are calculated by normalizing each subject's weights so that their sum is 1.00, as illustrated by the two matrices at the bottom left of Figure 14-46. This defines a set of r normalized weight vectors. The rth one is dropped (because it is a linear combination of the others) and the remaining ones are centered and normalized, as illustrated by the matrix at the bottom right in the figure. This is the complete definition of the flattening transformation used in the SPSS Multidimensional Scaling procedure. The rank of the flattened weights matrix is $r - 1$. Thus, any one of its variables is a linear combination of the remaining variables. Therefore, one of the variables can be dropped.

Figure 14-46
Flattened transformation (not drawn in SPSS)

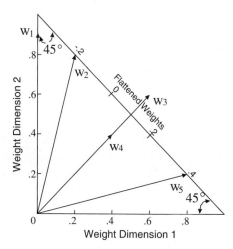

	Matrix Weights W				Normalized Matrix Weights W*			Flattened Weights
	I	II	Length		I	II	Sum	I
w_1	0	0.9	0.90	w_1*	0	1.0	1.0	–0.4
w_2	0.2	0.8	0.82	w_2*	0.2	0.8	1.0	–0.2
w_3	0.6	0.6	0.85	w_3*	0.5	0.5	1.0	0.1
w_4	0.4	0.4	0.56	w_4*	0.5	0.5	1.0	0.1
w_5	0.8	0.2	0.82	w_5*	0.8	0.2	1.0	0.4

The geometry underlying the idea of flattened weights, as implemented in SPSS, is also illustrated in Figure 14-46. The flattening transformation directly projects the weight vectors (labeled w_1 through w_5 in the figure) onto a space having one less dimension than the weight space and which is at a 45° angle to all dimensions. In the figure, the weight space (whose dimensions are the horizontal and vertical sides of the triangle) is two-dimensional, whereas the flattened weight space (which is the hypotenuse of the triangle) is one dimensional. Note that the weight vector w_3 extends beyond the hypotenuse before flattening, so its projection onto the flattened space is shorter than the original vector. However, the weight vectors w_1 and w_2 are lengthened in their projection onto the flattened space. Finally, we look at the points in the flattened space defined by the projections from the original weight space.

Notice that this flattening transformation does not map the angles in the weight space onto distances in the flattened space in a one-to-one fashion. (In fact, there is no one-to-one flattening transformation.) An angle that is at the edge of the weight space is represented by a distance that is larger than the distance for an equal angle in the middle of the weight space. That is, subjects who are peripheral in the weight space appear to be even more peripheral in the flattened space.

Bibliography

Agresti, A. 2002. *Categorical data analysis*. 2nd ed. New York: John Wiley and Sons.

Bartolucci, A. A., and M. D. Fraser. 1977. Comparative step-up and composite tests for selecting prognostic indicators associated with survival. *Biometrical Journal*, 19: 437–448.

Begg, C. B., and R. Gray. 1984. Calculations of polytomous logistic regression parameters using individualized regressions. *Biometrika*, 71: 11–18.

Belsley, D., E. Kuh, and R. Welsch. 1980. *Regression diagnostics: Identifying influential data and sources of collinearity*. New York: John Wiley and Sons.

Bowker, A. H., and G. K. Lieberman. 1972. *Engineering statistics*. 2nd ed. Englewood Cliffs, N.J.: Prentice Hall, Inc.

Carroll, J. D., and J. J. Chang. 1970. Analysis of individual differences in multidimensional scaling via an N-way generalization of Eckart-Young decomposition. *Psychometrika*, 35: 238–319.

Corbeil, R. R., and S. R. Searle. 1976. A comparison of variance component estimators. *Biometrics*, 32: 779–791.

Crowley, J., and B. E. Storer. 1983. Comment on "A reanalysis of the Stanford heart transplant data" by M. Aitkin, N. Laird, and B. Francis. *Journal of the American Statistical Association*, 78: 277–281.

Davis, J. A., and T. W. Smith. 1972–2002. General Social Survey (GSS). Chicago: National Opinion Research Center (NORC), University of Chicago.

DeCarlo, L.T. 2003. Using the PLUM procedure of SPSS to fit unequal variance and generalized signal detection models. Behavior Research Methods, Instruments & Computers, 2003, 35(1).

Doble, J., and J. Greene. 1999. Attitudes toward crime and punishment in Vermont: Public opinion about an experiment with restorative justice, 1999 (computer file). Englewood Cliffs, N.J.: Doble Research Associates, Inc. (producer), 2000. Ann Arbor, Mich.: Inter-university Consortium for Political and Social Research (distributor), 2001.

Draper, N. R., and H. Smith. 1998. An introduction to nonlinear estimation. In: *Applied regression analysis*, 3rd ed. New York: John Wiley and Sons.

Finney, D. J. 1971. *Probit analysis*. 3rd ed. Cambridge: Cambridge University Press.

Giesbrecht, F. G. 1983. An efficient procedure for computing MINQUE of variance components and generalized least squares estimates of fixed effect. *Communications in Statistics*, Part A, Theory and Methods, 12: 2169–2177.

Gill, P. E., W. M. Murray, M. A. Saunders, and M. H. Wright. 1986. User's guide for NPSOL (version 4.0): A FORTRAN package for nonlinear programming. Technical Report SOL 86–2, Department of Operations Research, Stanford University.

Haberman, S. J. 1982. Analysis of dispersion of multinomial responses. *Journal of the American Statistical Association*, 77: 568–580.

Hauck, W. W., and A. Donner. 1977. Wald's test as applied to hypotheses in logit analysis. *Journal of the American Statistical Association*, 72: 851–853.

Hedeker, D., R. D. Gibbons, and B. R. Flay. 1994. Random-effects regression models for clustered data with an example from smoking prevention research. *Journal of Consulting and Clinical Psychology*, 62: 4, 757–765.

Hemmerle, W. J., and H. O. Hartley. 1973. Computing maximum likelihood estimates for the mixed AOV model using the W-transformation. *Technometrics*, 15: 819–831.

Hocking, R. R. 1985. *The analysis of linear models*. Monterey, Calif.: Brooks/Cole Publishing.

Hosmer, D. W., and S. Lemeshow. 1999. *Applied survival analysis*. New York: John Wiley and Sons.

_____. 2000. *Applied logistic regression*. 2nd ed. New York: John Wiley and Sons.

Huynh, H., and L. S. Feldt. 1970. Conditions under which mean square ratios in repeated measurements designs have exact F-distributions. *Journal of the American Statistical Association*, 65: 1582–1589.

Jennrich, R. I., and P. F. Sampson. 1976. Newton-Raphson and related algorithms for maximum likelihood variance component estimation. *Technometrics*, 18: 11–17.

Jennrich, R. I., and M. D. Schluchter. 1986. Unbalanced repeated-measures models with structured covariance matrices. *Biometrics*, 42: 805–820.

Johnson, N. L., S. Kotz, and A. W. Kemp. 1992. *Univariate discrete distributions*. 2nd ed. New York: John Wiley and Sons.

Kalbfleisch, J. D., and R. L. Prentice. 1980. The statistical analysis of failure time data. New York: John Wiley and Sons.

Kaplan, E. L., and P. Meier. 1958. Nonparametric estimation from incomplete data *Journal of the American Statistical Association*, 53: 457–481.

Kelejian, H. H., and W. E. Oates. 1989. *Introduction to econometrics: Principles and applications*. New York: Harper and Row.

Klein, J. P., and M. L. Moeschberger. 1997. *Survival analysis: Techniques for censored and truncated data*. New York: Springer.

Kruskal, J. B. 1964. Nonmetric multidimensional scaling: A numerical method. *Psychometrika*, 29: 115–129.

LaMotte, L. R. 1973. Quadratic estimation of variance components. *Biometrics*, 29: 311–330.

Lawless, J. F., and K. Singhal. 1978. Efficient screening of nonnormal regression models. *Biometrics*, 34: 318–327.

MacCallum, R. C. 1977. Effects of conditionality on INDSCAL and ALSCAL weights. *Psychometrika*, 42: 297–305.

Magidson, Jay. 1981. Qualitative variance, entropy, and correlation ratios for nominal dependent variables. *Social Science Information*, 10: 177–194.

McCullagh, P., and J. A. Nelder. 1989. *Generalized linear models*. 2nd ed. London: Chapman and Hall.

McGee, V. E. 1968. Multidimensional scaling of N-sets of similarity measures. A nonmetric individual differences approach. *Multivariate Behavioral Research*, 3: 233–248.

Miller, P. W., and P. A. Volker. 1985. On the determination of occupational attainment and mobility. *Journal of Human Resources*, 20: 197–213.

Moré, J. J. 1977. The Levenberg-Marquardt algorithm: Implementation and theory in numerical analysis. In: *Lecture Notes in Mathematics* 630, G.A. Watson, ed. Berlin: Springer-Verlag. 105–116.

Norušis, M. J. 2005. *SPSS 14.0 Guide to Data Analysis*. Upper Saddle River, N.J.: Prentice Hall, Inc.

_____. 2005. *SPSS 14.0 Statistical Procedures Companion*. Upper Saddle River, N.J.: Prentice Hall, Inc.

Patterson, H. D., and R. Thompson. 1971. Recovery of interblock information when block sizes are unequal. *Biometrika,* 58: 545–554.

Pothoff, R. F., and S. N. Roy. 1964. A generalized multivariate analysis of variance model useful especially for growth curve problems. *Biometrika*, 51: 313–326.

Raftery, A. E. 1986. Choosing models for cross-classification. *American Sociological Review*, 51: 145–146.

Rao, C. R. 1973. *Linear statistical inference and its applications*. 2nd ed. New York: John Wiley and Sons.

Rao, C. R., and J. Kleffe. 1988. Estimation of variance components and applications. In: *North-Holland Series in Statistics and Probability* (Vol. 3). North-Holland, Amsterdam.

Raudenbush, S. W., and A. S. Bryk. 2002. Hierarchical linear models: Applications and data analysis methods. Thousand Oaks, Calif.: Sage Publications.

Schoenfeld, D. 1982. Partial residuals for the proportional hazards regression model. *Biometrika*, 69: 239–241.

Schwarz, G. 1978. Estimating the dimension of a model. *Annals of Statistics*, 6: 461–464.

Searle, S. R., G. Casella, and C. E. McCulloch. 1992. *Variance components*. New York: John Wiley and Sons.

Shepard, R. N. 1962. The analysis of proximities: Multidimensional scaling with an unknown distance function. *Psychometrika*, 21: 125–139, 219–246.

Singer, J. D. 1998. Using SAS PROC MIXED to fit multilevel models, hierarchical models, and individual growth models. *Journal of Educational and Behavioral Statistics*, 23: 4, 323–355.

Snedecor, G. W., and W. G. Cochran. 1980. *Statistical methods*. 7th ed. Ames, Ia.: Iowa State University Press.

Speed, F. M. 1979. Choice of sums of squares for estimation of components of variance. *Proceedings of Statistical Computing Section*, 55–58. Alexandria, Va.: American Statistical Association.

Stablein, D., W. Carter, and J. Novak. 1981. Analysis of survival data with non-proportional hazard functions. *Controlled Clinical Trials*, 2: 149–159.

Swets, J. A., W. P. Tanner, and T. G. Birdsall. 1961. Decision processes in perception. *Psychological Review*, 68.

Takane, Y., J. de Leeuw, and F. W. Young. 1977. Nonmetric individual differences multidimensional scaling: An alternating least squares method with optimal scaling features. *Psychometrika*, 42: 7–67.

Therneau, T., P. Grambsch, and T. Fleming. 1990. Martingale-based residuals for survival models. *Biometrika*, 77: 147–160.

Torgerson, W. S. 1952. Multidimensional scaling I: Theory and method. *Psychometrika*, 17: 401–419.

Willett, J. B. 1988. Questions and answers in the measurement of change. In: *Review of research in education (1988–1989)*, E. Rothkopf, ed. Washington, D.C.: American Educational Research Association.

Willett, J. B., and J. D. Singer. 2003. *Applied longitudinal data analysis: Modeling change and event occurrence*. Oxford: Oxford University Press.

Winer, B. J., R. Brown, and K. M. Michels. 1991. *Statistical principles in experimental design*. 3rd ed. New York: McGraw-Hill.

Young, F. W. 1974. Scaling replicated conditional rank order data. In: *Sociological Methodology*, D. Heise, ed. American Sociological Association, 129–170.

_____. 1981. Quantitative analysis of qualitative data. *Psychometrika*, 46: 357–388.

Young, F. W., and R. M. Hamer. 1987. *Multidimensional scaling: History, theory, and applications*. Hillsdale, N.J.: Lawrence Erlbaum Associates.

Young, F. W., and R. Lewyckyj. 1979. *ALSCAL-4 user's guide*. Carrboro, N.C.: Data Analysis and Theory Associates.

Index